S0-BJX-197

Mechanical Properties Of Work Materials

Edmund Isakov, Ph.D.

Hanser Gardner Publications
Cincinnati

Library of Congress Cataloging-in-Publication Data

Isakov, Edmund.
Mechanical properties of work materials / Edmund Isakov.
p. cm.
Includes bibliographical references and index.
ISBN 1-56990-294-1
1. Metals—Mechanical properties. I. Title

TA460 .I68 2000
671.5'3--dc21 99-088669

While the advice and information in *Mechanical Properties of Work Materials* are believed to be true, accurate, and reliable, neither the author nor the publisher can accept any legal responsibility for any errors, omissions, or damages that may arise out of the use of this advice and information. The author and publisher make no warranty of any kind, expressed or implied, with regard to the material contained in this work.

A ***Modern Machine Shop*** book published by
Gardner Publications, Metalworking's Premier Publisher
www.mmsonline.com

Hanser Gardner Publications
6915 Valley Avenue
Cincinnati, OH 45244-3029
www.hansergardner.com

1 2 3 4 5 6 05 04 03 02 01 00

Dedication

To my past: my mother, Vera; my father Isaac; my brother Sam.

To my present: my wife, Yelena; my daughter, Marina; my sister Kira.

To my future: my grandson, Andy; my granddaughter, Sasha.

Acknowledgments

The author would like to thank the following people for their help, encouragement, and also for stimulating discussions on specific subjects.

Kennametal Inc.	Mark Huston, Bill Kennedy, John Hale, Vickie Plesko
PromaTec Systems Inc.	Alan Lyons

Table Of Contents

1. Introduction 7

2. Classification of Work Materials 7
- 2.1 Ferrous Alloys 7
  - 2.1.1 Carbon Steels 7
  - 2.1.2 Low-Alloy Steels 8
  - 2.1.3 High-Alloy Steels 8
  - 2.1.4 Cast Irons 9
- 2.2 Nonferrous Alloys 10
  - 2.2.1 Aluminum Alloys 10
  - 2.2.2 Titanium Alloys 11
  - 2.2.3 Nickel-Base Alloys 12
  - 2.2.4 Cobalt-Base Alloys 12
  - 2.2.5 Copper Alloys 13

3. Hardness 14
- 3.1 Rockwell Hardness 14
- 3.2 Brinell Hardness 15
- 3.3 Vickers Hardness 15
- 3.4 Knopp Hardness 15
- 3.5 Scleroscope Hardness 16

4. Hardness-To-Hardness Conversion 16

5. Strength-Hardness Relationship Of Work Materials 19

6. Concluding Remarks 24

Appendix I. Conversion Of Hardness Numbers
1. Linear Regression Formulas for Hardness Conversion 29
2. Conversion of Rockwell B Hardness Numbers into Brinell Hardness Numbers 30
3. Conversion of Rockwell C Hardness Numbers into Brinell Hardness Numbers 37
4. Conversion of Vickers Hardness Numbers into Brinell Hardness Numbers 46
5. Conversion of Knoop Hardness Numbers into Brinell Hardness Numbers 53
6. Conversion of Scleroscope Hardness Numbers into Brinell Hardness Numbers 62

Appendix II. Statistical Treatment Of Data 69

Appendix III. Hardness And Tensile Strength Relationship For Carbon Steels 75

Appendix IV. Hardness And Tensile Strength Relationship For Low-Alloy Steels 97

Appendix V. Hardness And Tensile Strength Relationship For Stainless Steels 119

Appendix VI. Hardness And Tensile Strength Relationship For Tool Steels 135

Appendix VII. Hardness And Tensile Strength Relationship For Cast Irons 149

Appendix VIII. Hardness And Tensile Strength Relationship For Titanium Alloys And Pure Titanium 159

Appendix IX. Hardness And Tensile Strength Relationship For Wrought Aluminum Alloys 169

Appendix X. Hardness And Tensile Strength Relationship For Cast Aluminum Alloys 193

Appendix XI. Hardness And Tensile Strength Relationship For Wrought Copper Alloys 201

Appendix XII. Hardness And Tensile Strength Relationship For Cast Copper Alloys 217

References 225

1. Introduction

Three elements define metalcutting as an integrated system: 1) a wide variety of work materials; 2) cutting tools; and 3) machine tools. The work material properties should always be taken into consideration with the other two elements to make the metalcutting system *effective.*

Work materials, the majority of which are metals, differ from each other by mechanical and physical properties. Among mechanical properties, the hardness and strength are essential since they affect the cutting forces and power consumption.

Traditionally, guidelines for the selection of cutting tool geometries and grades have been based on *tool application considerations* with only *limited attention to workpiece characteristics* such as the type of material and its hardness number range, which is sometimes beyond measurable values.

Appropriate classification of work materials, combined with knowledge of their mechanical properties, and the relationship between different materials, enhances the guidelines for the application of cutting tools and produces higher accuracy when calculating cutting forces and machining power consumption.

2. Classification Of Work Materials

The majority of industrial applications of machining are with metals. All metals that are used as workpiece materials can be characterized as ferrous alloys and nonferrous alloys.

Ferrous alloys are all steels and cast irons. Alloys of iron (Fe) and carbon (C) that contain up to 2% C are classified as steels, while those containing over 2% C are classified as cast irons.

Nonferrous alloys are alloys that contain metals other than iron. Some nonferrous alloys contain iron as an alloying element only.

2.1 Ferrous Alloys

The most common ferrous alloys are carbon steels, low-alloy steels, high-alloy steels, and cast irons.

The American Iron and Steel Institute (AISI) and the Society of Automotive Engineers (SAE) use an identical four-digit number to designate carbon and low-alloy steels.

2.1.1 Carbon Steels

Carbon steels are designated as 10xx, 11xx, 12xx, and 15xx. The initial two digits relate to sulfur (S), phosphorus (P), and manganese (Mn) content:

- 10xx – nonresulfurized grades (0.050% max S; 0.040% max P; 1.00% max Mn)
- 11xx – resulfurized grades (0.33% max S; 0.040% max P; 1.65% max Mn)

- 12xx – resulfurized and rephosphorized grades (0.35% max S; 0.12% max P; 1.00% max Mn)
- 15xx – nonresulfurized grades (0.050% max S; 0.040% max P; more than 1.00% Mn).

The last two digits indicate middle of the carbon range in hundredths of a percent.

There are three carbon steel groups:

- Low-carbon steels contain from 0.06% to 0.28% C: AISI 1005 to 1026, AISI 1108 to 1119, all AISI 12xx, and AISI 1513 to 1527
- Medium-carbon steels containing from 0.25% to 0.55% C: AISI 1029 to 1053, AISI 1137 to 1151, and AISI 1541 to 1552
- High-carbon steels containing from 0.50% to 1.00% C: AISI 1055 to 1095 and AISI 1561 to 1566.

The 11xx series are free-machining grades. The 12xx series are also free-machining grades, which provide increased chip control. The 15xx series grades contain from 1.05% to 1.65% Mn for greater hardenability.

2.1.2 Low-Alloy Steels

Low-alloy steels contain more than one of the following alloying elements: manganese (Mn), silicon (Si), nickel (Ni), chromium (Cr), molybdenum (Mo), and vanadium (V). Total content of alloying elements does not exceed 8%. Low-alloy steels are designated by the AISI four-digit numerical code similar to carbon steels. The first two digits in the code indicate the major alloying element or elements:

- 13xx — Manganese steels
- 23xx, 25xx — Nickel steels
- 31xx, 32xx, 33xx, 34xx — Nickel-chromium steels
- 40xx, 44xx — Molybdenum steels
- 41xx — Chromium-molybdenum steels
- 43xx, 47xx, 8xxx, 93xx, 94xx, 97xx, 98xx — Nickel-chromium-molybdenum steels
- 46xx, 48xx — Nickel-molybdenum steels
- 50xx, 51xx, 50xxx, 51xxx, 52xxx — Chromium steels
- 61xx — Chromium-vanadium steels
- 92xx — Silicon-manganese steels

2.1.3 High-Alloy Steels

High-alloy steels contain more than 8% of alloying elements. Typical high-alloy steels are wrought stainless, cast stainless, and tool steels.

Wrought stainless steels are divided into four general classes:

- Austenitic types of the chromium-nickel-manganese 200-series and the chromium-nickel 300-series
- Martensitic types of the straight-chromium, hardenable 400-series

- Ferritic types of the straight-chromium, nonhardenable 400-series
- Precipitation-hardening types of chromium-nickel alloys with additional elements that are hardenable by solution treating and aging (0.050% max S; 0.040% max P).

Cast stainless steels are classified as corrosion-resistant and heat-resistant alloys. *Corrosion-resistant compositions (C-Type)* are used in the environments below 1200°F. C-Type alloys are divided into three groups:

- Chromium steels (11.5-30.0% Cr and 1.0-4.0% Ni)
- Chromium-nickel steels (10.5-30.0% Cr and 3.5-22.0% Ni)
- Nickel-chromium steels (22.0-34.0% Ni and 18.0-22.0% Cr).

Heat-resistant compositions (H-type) are capable of withstanding operating temperatures in excess of 1200°F. There are three basic categories of H-type alloys:

- Iron-chromium steels (10 to 30% Cr and little or no nickel)
- Iron-chromium-nickel steels (more than 13% Cr and more than 7% Ni, always more chromium than nickel)
- Iron-nickel-chromium steels (more than 25% Ni and more than 10% Cr, always more nickel than chromium).

Tool steels are divided into the following categories, each with an identifying letter symbol:

- Water-hardening tool steels, W
- Shock-resisting tool steels, S
- Cold work tool steels
  - Oil-hardening types, O
  - Air-hardening types, A
  - High-carbon, high-chromium types, D
- Hot work tool steels, H
- Special-purpose tool steels
  - Low-alloy types, L
  - Carbon-tungsten types, F
  - Mold steels, P
- High speed tool steels
  - Tungsten-base types, T
  - Molybdenum-base types, M.

Note: Many types of tool steels are consumed in large quantities for nontool uses.

2.1.4 Cast Irons

Cast irons are classified as common and special cast irons. Common cast irons contain iron (Fe), carbon (C), silicon (Si), and manganese (Mn) as major alloying elements, and some minor alloying elements such as chromium (Cr), nickel (Ni), molybdenum (Mo), and copper (Cu).

Common cast irons are:

- Gray (2.5-4.0% C, 1.0-3.0% Si, 0.2-1.0% Mn, for example, SAE grades: G1800, G2500, G3000, G3500, and G4000)
- Ductile (3.0-4.0% C, 1.8-2.8% Si, 0.1-1.0% Mn, for example, SAE grades: D4018, D4512, D5506, and D7003; ASTM grades: 60-40-18, 65-45-12, 80-55-06, 100-70-03, and 120-90-02)
- Malleable (2.2-2.9% C, 0.9-1.9% Si, 0.2-1.3% Mn, for example, ASTM and SAE grades: M3210, M4504, M5003, M5503, M7002, and M8501).

Special or alloy cast irons are based on the iron-carbon-silicon system with alloying elements such as chromium, nickel, molybdenum, and manganese.

Special or alloy cast irons are:

- Abrasion-resistant white irons (2.0-3.7% C, 0.2-2.0% Mn, 0.3-2.2% Si, 1.0-28.0% Cr, 1.5-5.0% Ni, 0.5-3.5% Mo, 0.0-1.2% Cu)
- Corrosion-resistant irons (3.0-4.0% C, 0.3-4.5% Mn, 0.5-17.0% Si, 1.0-35% Cr, 5.0-36% Ni, 1.0-4.0% Mo, 0.5-7.5% Cu)
- Heat-resistant gray irons (2.0-3.0% C, 0.4-1.5% Mn, 1.0-7.0% Si, 1.8-6.0% Cr, 13.0-43% Ni, 1.0% Mo, 7.5-10.0% Cu)
- Heat-resistant ductile irons (2.0-3.8% C, 0.2-2.4% Mn, 0.5-6% Si, 1.8-35% Cr, 1.5-36% Ni, 1.0-2.0% Mo).

2.2 Nonferrous Alloys

The most common nonferrous work materials are aluminum alloys, titanium alloys, nickel-base alloys, cobalt-base alloys, and copper alloys.

2.2.1 Aluminum Alloys

Aluminum alloys are divided into two categories: wrought and casting compositions.

Wrought aluminum alloys are identified by a four-digit numerical code in which the first digit indicates the major alloying element:

- 2xxx alloys in which copper is the principal alloying element (1.0-6.8% Cu)
- 3xxx alloys in which manganese is the principal alloying element (0.2-1.8% Mn)
- 4xxx alloys in which silicon is the principal alloying element (1.0-13.5% Si)
- 5xxx alloys in which magnesium is the principal alloying element (0.5-5.6% Mg)
- 6xxx alloys in which magnesium and silicon are principal alloying elements (0.3-1.5% Mg and 0.2-1.8% Si)
- 7xxx alloys in which zinc is the principal alloying element, but other elements such as copper, magnesium, chromium, and zirconium (Zr) may be specified (1.3-8.7% Zr).

Casting aluminum alloys are identified by a three-digit numerical code followed by a decimal value. The first digit indicates the major alloying element and the last digit, which is to the right of the decimal point, indicates the product form: xxx.0 is for casting, xxx.1 is for ingot. The groups of casting aluminum alloys are as follows:

- 2xx.x alloys in which copper is the principal alloying element (3.5–10.7% Cu)
- 3xx.x alloys in which silicon and copper are the principal alloying elements (4.5–23.0% Si and 0.5–5.0% Cu)
- 4xx.x alloys in which silicon is the principal alloying element (4.5–13.0% Si)
- 5xx.x alloys in which magnesium is the principal alloying element (2.5–10.6% Mg)
- 7xx.x alloys in which zinc is the principal alloying element (2.7–8.0% Zn)
- 8xx.x alloys in which tin is the principal alloying element (5.5–7.0% Sn).

2.2.2 Titanium Alloys

Titanium alloys are classified as alpha alloys, alpha-beta alloys, and beta alloys.

Alpha alloys are slightly less corrosion resistant but higher in strength than unalloyed titanium. Alpha alloys are ductile and cannot be strengthened by heat treatment. The principal alloying element in alpha titanium alloys is aluminum (Al). Other alloying elements are tin (Sn), zirconium (Zr), molybdenum (Mo), vanadium (V), tantalum (Ta), niobium (Nb), and silicon (Si). The number of alloying elements and their composition depend on the applications.

Typical applications of some alpha titanium alloy grades:

- Ti-8Al-1V-1Mo (8% Al, 1% V, 1% Mo) – fan blades, jet engine components
- Ti-5Al-2.5Sn (4-6% Al, 2-3% Sn) – aerospace structural members
- Ti-6Al-2Nb-1Ta-0.8Mo (6% Al, 2% Nb, 1% Ta, 0.8% Mo) – pressure vessels.

Alpha-beta alloys can be higher in tensile strength than alpha alloys due to heat treatment. The principal alloying elements in alpha-beta alloys are aluminum (Al) and vanadium (V). Other alloying elements are tin (Sn), zirconium (Zr), molybdenum (Mo), chromium (Cr), and manganese (Mn). The number of alloying elements and their composition depend on the applications.

Typical applications of some alpha-beta titanium alloy grades:

- Ti-6Al-4V (5.5-6.75% Al, 3.5-4.5% V) – airframe components, aircraft gas turbine blades

- Ti-6Al-6V-2Sn (6% Al, 6% V, 2% Sn) – parts requiring high strength at and above 600°F
- Ti-3Al-2.5V (2.5-3.5% Al, 2.0-3.0% V) – seamless tubing for aircraft hydraulic parts.

Beta alloys provide increased fracture toughness at a given strength level in comparison with alpha-beta alloys. Major alloying elements in beta alloys are vanadium (V), chromium (Cr), and aluminum (Al).

Typical applications of some beta titanium alloy grades:

- Ti-13V-11Cr-3Al (12.5-14.5% V, 10-12% Cr, 2.5-3.5% Al) – high-strength airframe components
- Ti-8Mo-8V-2Fe-3Al (7.5-8.5% Mo, 7.5-8.5% V, 1.6-2.4% Fe, 2.6-3.4% Al) – forgings for aerospace structures
- Ti-10V-2Fe-3Al (9.2-11.8% V, 1.6-2.5% Fe, 2.5-3.5% Al) – high-strength airframe components.

2.2.3 Nickel-Base Alloys

Nickel-base alloys are used for a wide variety of applications, which involve heat resistance and corrosion resistance. Nickel-base alloys contain more than a dozen alloying elements, two of which are the principal alloying elements: nickel (30 to 78%) and chromium (14 to 31%). There are two major groups of nickel-base alloys:

1. Nickel-chromium and nickel-chromium-iron alloys.
2. Iron-nickel-chromium alloys.

The following work materials represent the first group: Alloys 600, 601, 617, 625, 690, 718, X750, 751, MA754, C-22, C-276, G3, HX, S, W, and X.

The following work materials are from the second group: Alloys 556, 800, 800HT, 825, 925, 20Cb3, 20Mo-4, and 20Mo-6.

2.2.4 Cobalt-Base Alloys

Cobalt-base alloys are used for wear resistant, heat resistant, and corrosion resistant applications. Cobalt-base alloys contain about ten alloying elements, two of which are the principal alloying elements: cobalt (32 to 62%) and chromium (18 to 31%). All cobalt-base alloys are divided into the following groups:

1. Wear-resistant alloys
2. High-temperature alloys
3. Corrosion-resistant alloys.

Wear-resistant alloys are Stellites 1, 4, 6, 6K, 12, 21, 190, 306, F, Haynes alloy 6B, Tribaloy T-800.

High-temperature alloys are Haynes alloy 25, Haynes alloy 188, and MAR-M alloy 509.

Corrosion-resistant alloys are MP35N alloy and Haynes alloy 1233.

2.2.5 Copper Alloys

Copper alloys are divided into five groups. Each of the groups contains one of five major alloying elements as the primary alloying ingredient:

- Brasses in which zinc is the principal alloying element
- Phosphor bronzes in which tin is the principal alloying element
- Aluminum bronzes in which aluminum is the principal alloying element
- Silicon bronzes in which silicon is the principal alloying element
- Copper-nickels and nickel silvers in which nickel is the principal alloying element.

The UNS (Unified Numbering System) numbers consisting of the letter C and a five-digit numerical code are used to classify *wrought* and *cast copper alloys*. The first digit for wrought copper alloys is any number from 1 to 7, and it is 8 or 9 for cast copper alloys. The first digit of the numerical code also indicates the group of copper alloys and the alloying elements.

Wrought copper alloys:

- C16200-C19600 are high-copper alloys, alloying element is copper (>96%)
- C20500-C28580 are brasses, alloying element is zinc (5-40%)
- C31200-C38590 are leaded brasses, alloying elements are zinc (8.1-40.0%), lead (0.3-3.0%)
- C40400-C49080 are tin brasses, alloying elements are zinc (3.0-39.0%) and tin (0.5-2.2%)
- C50100-C52400 are phosphor bronzes, alloying element is tin (1.2-10.0%)
- C53200-C54800 are leaded phosphor bronzes, alloying elements are tin (4%), lead (4%), and zinc (4%)
- C60600-C64400 are aluminum bronzes, major alloying element is aluminum (2.8-13.0%), some aluminum bronzes contain iron (1.0-4.3%), nickel (2.0-5.0%), and silicon (1.0-1.8%)
- C64700-C66100 are silicon bronzes, principal alloying element is silicon (1.5-3.0%)
- C66400-C69900 are copper-zinc alloys, major alloying element is zinc (14.5-39.0%), other alloying elements are aluminum, manganese, tin, and silicon
- C70000-C72500 are copper-nickels, principal alloying element is nickel (3-31%)
- C73200-C79900 are nickel silvers, alloying elements are nickel (8-18%) and zinc (10-29%).

Cast copper alloys:

- C81300-C82800 are high-copper alloys, alloying element is copper (>94%)
- C83300-C84800 are red and leaded red brasses, alloying elements are zinc (4.0-15.0%), lead (1.5-7.0%), and tin (1.5-5.0%)

- C85200-C85800 are yellow and yellow leaded brasses, alloying elements are zinc (24-40%), lead (1-3%), and tin (1%)
- C86100-C86800 are manganese bronzes and leaded manganese bronzes, principal alloying elements are zinc (21-41%) and manganese (3-4%)
- C87300-C87900 are silicon bronzes and silicon brasses, alloying elements are zinc (5-34%) and silicon (3.0-4.5%)
- C90200-C91700 are tin bronzes, principal alloying element is tin (7-19%)
- C92200-C92900 are leaded tin bronzes, principal alloying elements are tin (6-16%) and lead (1-5%)
- C93200-C94500 are high-leaded tin bronzes, principal alloying elements are lead (7-25%) and tin (5-13%)
- C94700-C94900 are nickel-tin bronzes, principal alloying elements are nickel (5%) and tin (5%)
- C95200-C95800 are aluminum bronzes, principal alloying element is aluminum (7-11%)
- C96200-C96700 are copper-nickels, principal alloying element is nickel (20-30%)
- C97300-C97800 are nickel silvers, principal alloying elements are nickel (12-20%) and zinc (2-20%)
- C98200-C98800 are leaded coppers, alloying elements are lead (24.0-40.0%) and silver (1.5-5.5%).

3. Hardness

Hardness of metals is a measure of their resistance to deformation. From the machining viewpoint, hardness is the resistance to cutting. Various static indentation and dynamic methods are used to measure hardness. Typical static indentation methods are Rockwell, Brinell, Vickers, and Knoop hardness testing, which use different loads and indenters. A typical dynamic (rebound) method is Scleroscope hardness testing, also known as the Shore method.

3.1 Rockwell Hardness

The Rockwell hardness value is a combination of a number, the symbol **HR** and the scale designation. For example, **80 HRB** represents the Rockwell hardness number of **80** on the Rockwell **B** scale and **40 HRC** represent the Rockwell hardness number of **40** on the Rockwell **C** scale. These two scales are used to measure hardness of a wide variety of metals and alloys.

Scale B (100-kgf load and the 1/16-inch ball indenter) is used to measure hardness of nonferrous metals and alloys, soft grades of cast irons, and unhardened steels. Readings between approximately **20** (minimum value) and **100** (maximum value) are normal ranges. The Rockwell B hardness numbers for carbon and alloy steels in the annealed, normalized, and quenched-and-tempered conditions are between **55** and **100.**

Scale C (150-kgf load and the 120º sphero-conical diamond indenter) is used to measure hardness of steels, cast irons, titanium, titanium alloys, and other metals harder than **100 HRB**. Readings between approximately **20** (the minimum value) and **68** (the maximum value) are most accurate and accepted as normal range.

Scale A (60-kgf load and the same indenter as for the scale C) is limited to thin steel and shallow casehardened steel.

3.2 Brinell Hardness

The Brinell hardness value is a combination of a number and the symbol **HB**. For example, **300 HB** represents the Brinell hardness number of **300**. Major test loads are 500- and 3000-kgf, and indenters are standard 10-mm-diameter balls.

The 500-kgf load and hardened steel ball are used for testing copper and aluminum alloys, and soft steels. The most accurate readings are between **16 HB** (very soft aluminum) and **100 HB** (unhardened and soft-tempered steels, gray and malleable cast irons, nonferrous alloys).

The 3000-kgf load and hardened steel ball are used for testing steels, cast irons, titanium alloys, nickel-base alloys, and cobalt-base alloys. The most accurate readings are between **100 HB** and **444 HB. A tungsten carbide ball** indenter is recommended for similar materials with the hardness number range between **444 HB** and **627 HB**.

3.3 Vickers Hardness

The Vickers hardness value is a combination of a number and the symbol **HV**. For example, **400 HV** represents the Vickers hardness number of **400**. The numerical value equals the applied load in kilograms divided by the surface area of the indentation in square millimeters. The Vickers indenter is a square-based pyramidal diamond having an included angle of 136°. The load, which is applied for 30 seconds, may either be 5, 10, 20, 30, 50 or 120 kilograms. The most commonly used load is 50-kgf.

3.4 Knoop Hardness

The Knoop hardness value is a combination of a number and the symbol **HK**. Knoop hardness numbers are similar to Vickers hardness numbers, since both test methods indicate the stress values expressed as kilograms per square millimeter (kgf/mm^2). The Knoop indenter is a rhombic-based pyramidal diamond with an included longitudinal angle of 172° 30' and an included transverse angle of 130°. The load may either be 100, 500, 1000 or 3000 grams.

Vickers and **Knoop** hardness tests are often used for measuring hardness close to an edge of a workpiece when Rockwell or Brinell hardness tests cannot be employed because their indenters require significantly higher loads. Vickers and Knoop hardness tests are applicable to thin metals, hard and brittle materials, and very shallow carburized or nitrided surfaces. For nonhomogeneous materials, such as cast irons and powder metallurgy parts (a mixture of both soft and very hard microconstituents), Vickers and Knoop testing are the only methods to accurately determine microhardness.

3.5 Scleroscope Hardness

The Scleroscope hardness number is a measure of the height of rebound of a diamond-tipped hammer falling from approximately 10 inches onto the surface of the metal being tested. The hammer weighs about 40 grains (2.6 grams). The harder the metal, the greater the rebound. The Scleroscope scale is graduated from 0 to 140 units. Hard steel tests about 100, medium-hard about 50, and soft metals 10 to 15 units. Readings between **17** (minimum value) and **97** (maximum value) are most accurate and accepted as normal range. Scleroscope hardness tests are used when numerous quality checks must be made or for very large specimens such as forged steel, wrought alloy steel rolls, and cast iron.

More information about hardness testing equipment, testing procedures, limitations, and hardness conversion tables can be found in various publications (Ref. 1, Ref. 2, p. 523-530).

4. Hardness-To-Hardness Conversion

Equivalent hardness numbers (HRB, HRC, HB, HV, and HK) for metals can be found in various handbooks. However, from a practical standpoint it is important to be able to convert one type of hardness number into another, especially when handbooks are not available. In order to express relationship between any two types of hardness numbers (let say, x and y) by a simple equation, the method of linear regression was used.

To perform linear regression analysis, the hardness numbers were set by groups of 5 to 20 data points in each group. Statistical treatment of data in each group produced regression equations of the straight lines with the correlation coefficients $r_{xy} > 0.90$ indicating a strong linear relationship between x and y variables:

$$y = ax \pm b$$

where x and y are any two types of hardness numbers,
a and b are parameters of a straight line, a is a slope and b is an intercept.

The linear regression formulas for slope, intercept, and the correlation coefficient (Ref. 3) are placed in Appendix I. Also, there are tables showing how linear regression analysis was performed to develop hardness-to-hardness conversion equations.

Hardness conversion equations are listed in Tables 1 to 5. Equations in Table 1 and Table 2 were developed to convert Rockwell hardness numbers (B and C scales) into Brinell hardness numbers.

Table 1. Brinell - Rockwell B Hardness Relationship

Rockwell B Hardness Numbers (HRB)		Equations to Convert Rockwell B Hardness (HRB) into Brinell Hardness (HB)
from	to	
41	54	***HB = 0.897 HRB + 38.3****
55	59	***HB = 1.500 HRB + 17.3***
60	69	***HB = 1.800 HRB – 1.4***
70	79	***HB = 2.394 HRB – 42.7***
80	89	***HB = 3.297 HRB – 114.4***
90	95	***HB = 5.000 HRB – 265.0***
96	100	***HB = 6.600 HRB – 418.0***

*Brinell hardness numbers (HB) obtained by this equation are the same as those being measured at 500-kgf load. Brinell hardness numbers calculated by other equations are the same as those measured at 3000-kgf load.

Table 2. Brinell - Rockwell C Hardness Relationship

Rockwell C Hardness Numbers (HRC)		Equations to Convert Rockwell C Hardness (HRC) into Brinell Hardness (HB)
from	to	
20	25	***HB = 5.329 HRC + 119.6***
26	30	***HB = 6.984 HRC + 76.1***
31	35	***HB = 8.379 HRC + 33.7***
36	40	***HB = 8.872 HRC + 16.0***
41	45	***HB = 10.025 HRC – 30.8***
46	50	***HB = 12.473 HRC – 142.6***
51	55	***HB = 15.962 HRC – 318.7***
56	60	***HB = 19.038 HRC – 489.4***
61	66	***HB = 17.602 HRC – 403.8***

Equations in Tables 3, 4, and 5 were developed to convert Vickers, Knoop, and Scleroscope hardness numbers into Brinell hardness numbers respectively.

Table 3. Brinell - Vickers Hardness Relationship

Vickers Hardness Numbers (HV)		Equations to Convert Vickers Hardness (HV) into Brinell Hardness (HB)
from	to	
85	149	***HB = 0.959 HV − 0.8***
150	199	***HB = 0.949 HV + 0.9***
200	249	***HB = 0.954 HV − 0.7***
250	299	***HB = 0.922 HV + 7.3***
300	399	***HB = 0.944 HV + 1.2***
400	499	***HB = 0.909 HV + 15.1***
500	670	***HB = 0.940 HV − 0.2***

Table 4. Brinell - Knoop Hardness Relationship

Knoop Hardness Numbers (HK)		Equations to Convert Knoop Hardness (HK) into Brinell Hardness (HB)
from	to	
100	149	***HB = 0.927 HK − 9.0***
150	199	***HB = 0.901 HK − 4.6***
200	249	***HB = 0.994 HK − 23.5***
250	299	***HB = 0.952 HK − 11.5***
300	349	***HB = 1.030 HK − 34.3***
350	399	***HB = 0.871 HK + 21.5***
400	499	***HB = 0.774 HK + 60.0***
500	599	***HB = 0.841 HK + 26.2***
600	700	***HB = 0.919 HK − 20.2***

Table 5. Brinell - Scleroscope Hardness Relationship

Scleroscope Hardness Numbers (HS)		Equations to Convert Scleroscope Hardness (HS) into Brinell Hardness (HB)
from	to	
17	29	***HB = 7.000 HS − 11.5***
30	39	***HB = 7.006 HS − 11.5***
40	49	***HB = 7.428 HS − 27.6***
50	59	***HB = 7.771 HS − 45.3***
60	69	***HB = 9.750 HS − 167.3***
70	79	***HB = 12.321 HS − 346.4***

5. Stength-Hardness Relationship Of Work Materials

When the cutting edge penetrates into a work material at a given depth of cut and feed rate, a deformed layer from the workpiece flows as a chip. The mechanism of chip formation is a complex process of elastic-plastic deformation of metal removed from the workpiece. This process can be simplified by three phases taking place during cutting.

1. Material in the "cutting zone" (between the cutting edge and the undeformed layer of work material) is always under *compression*.
2. The deformed layer in the "cutting zone" transforms into a chip when a cutting edge generates such stress that the *ultimate tensile strength* or *compressive strength* of the work material is overcome.
3. Separation of the chip from the parent material takes place in the "shear zone" near the cutting edge due to the *shear stress*.

Major properties of the work materials that influence the choice of a cutting insert grade, geometry, and machining conditions are the **hardness** and the **ultimate tensile strength**. Steels that have been "fully hardened" to the same hardness when quenched will have the same tensile and yield strengths regardless of composition and alloying elements (Ref. 2, p. 477). The same publication recommends two formulas (p. 530) for approximate relationship between Brinell hardness (**HB**) and tensile strength (**σ**, psi) that can be applied to steels.

$$\sigma - 515\,HB, \text{ if } HB \leq 175 \quad (1)$$

$$\sigma - 490\,HB, \text{ if } HB > 175 \quad (2)$$

However, these formulas provide the accuracy of ± 5% or worse when calculating the ultimate tensile strength of carbon and low-alloy steels. If the same formulas are used for stainless and tool steels, the accuracy is ± 10% or worse.

In order to develop more accurate formulas, the author applied statistical treatment of data and linear regression method to analyze the hardness-strength relationship. The following publications provide hardness-strength data for carbon, low-alloy, stainless, and tool steels (Ref. 4, 5, 6); for gray, ductile, and malleable cast irons (Ref. 5); for aluminum alloys, copper alloys (Ref. 7); for titanium and titanium alloys (Ref. 7, 8). These analyses resulted in the development of two types of formulas to calculate the ultimate tensile strength as a function of the work material hardness expressed primarily in Brinell and Rockwell numbers. The first type of formula is based on the statistical treatment of data (details of the statistical treatment of data are described in Appendix II). The second type of formula is based on the linear regression method described earlier (Appendix I). The first type of formula can be used to calculate the ultimate tensile strength of carbon steels, low-alloy steels, some stainless and tool steels, some ductile and malleable cast irons, some titanium alloys, and aluminum alloys. For example:

$$\sigma = \mathbf{500\ HB}\ \text{(psi)} \quad (3)$$

Formula (3) is recommended for calculating ultimate tensile strength of low-carbon, medium-carbon, high-carbon steels, low-alloy steels, and some stainless steels. The range of Brinell hardness numbers for these steels is:

- (90-250) HB for low-carbon steels
- (130-270) HB for medium-carbon steels
- (170-360) HB for high-carbon steels
- (140-370) HB for most low-alloy steels including (200-580) HB for AISI 4140 steel
- (155-595) HB for martensitic 400-series stainless steels
- (250-430) HB for precipitation-hardening stainless steels (PH and AM types, Custom 450).

The exceptions were found for AISI 4340 low-alloy steel with the hardness range of (390-600) HB and for precipitation-hardening stainless steels: 17-7 PH, PH 15-7 Mo (HB > 430), and Custom 455 with the hardness range of (370-440) HB. Formula (4) and formula (5) should be used to calculate the ultimate tensile strength of AISI 4340 steel and precipitation-hardening stainless steels respectively:

$$\sigma = \mathbf{540\ HB}\ \text{(psi)} \quad (4)$$
$$\sigma = \mathbf{550\ HB}\ \text{(psi)} \quad (5)$$

Formula (6) was developed for calculating the ultimate tensile strength of the austenitic 300-series stainless steels, and formula (7) can be used to calculate ultimate tensile strength of the austenitic 300-series and the ferritic 400-series stainless steels:

$$\sigma = \mathbf{560\ HB}\ \text{(psi)} \quad (6)$$
$$\sigma = \mathbf{470\ HB}\ \text{(psi)} \quad (7)$$

Formula (6) provides the accuracy of ± 7% or better if used for the hardness range of (140-180) HB. Formula (7) provides the accuracy of ± 7% or better when used for 300-series stainless steels with the hardness range of (190-370) HB. The same formula provides the accuracy of ± 3% or better when applied to the ferritic 400-series stainless steels with the hardness range of (140-190) HB.

To convert ultimate tensile strength from **psi ($lb/in.^2$)** into **N/mm^2** (metric unit of measure), the value of $\boldsymbol{\sigma}$ must be divided by the conversion factor of **145.0377**. For example,

$$\sigma = \mathbf{500\ HB}\ \text{(psi)} = \mathbf{3.447\ HB}\ (\text{N/mm}^2) \quad (8)$$

Statistical treatment of hardness-strength data for a variety of work materials is given in Appendixes III to XII. Type of work materials, hardness range, and formulas for ultimate tensile strength are summarized in Tables 6-13.

Table 6. Ultimate Strength – Hardness Relationship for Stainless Steels

Class	Designation and grade	Hardness range	Ultimate tensile strength
Austenitic	AISI Type 201	(250 - 400) HB	σ = 606 HB – 31660
Austenitic	AISI 300-series	(140 - 180) HB	σ = 457 HB + 16910
Austenitic	AISI 300-series	(190 - 370) HB	σ = 534 HB – 16280
Ferritic	AISI 400-series	(140 - 190) HB	σ = 430 HB + 6530
Martensitic	AISI 400-series	(156 - 595) HB	σ = 508 HB – 3900
Precipitation-hardening	PH, AM, Custom series	(250 - 430) HB	σ = 557 HB – 20710
Precipitation-hardening	PH, AM, Custom series	(25 - 46) HRC	σ = 4725 HRC – 2370

Table 7. Ultimate Strength – Hardness Relationship for Tool Steels

Categories	Designation and grade	Hardness range	Ultimate tensile strength
Shock – resisting	AISI Type S1	(42 - 58) HRC	σ = 7014 HRC – 94200
Shock – resisting	AISI Type S5	(37 - 59) HRC	σ = 7829 HRC – 131700
Shock – resisting	AISI Type S7	(39 - 58) HRC	σ = 6923 HRC – 92700
Oil – hardening	AISI Type O1	(31 - 50) HRC	σ = 5899 HRC – 60120
Oil – hardening	AISI Type O7	(31 - 54) HRC	σ = 6332 HRC – 78970
Air – hardening	AISI Type A2	(30 - 60) HRC	σ = 5920 HRC – 53270
Chromium hot-work	AISI Types H11 and H13	(30 - 57) HRC	σ = 6387 HRC – 62850
Special – purpose	AISI Type L2	(30 - 54) HRC	σ = 6227 HRC – 59980
Special – purpose	AISI Type L6	(32 - 54) HRC	σ = 6843 HRC – 83900
Mold	AISI Type P20	(26 - 54) HRC	σ = 5535 HRC – 24880
Mold	AISI Type P20	(260 - 540) HB	σ = 545 HB – 11770
Mold	AISI Type P20	(260 - 540) HB	σ = 520 HB

Table 8. Ultimate Strength – Hardness Relationship for Cast Irons

Type of Cast Iron	Designation	Hardness range	Ultimate tensile strength
Gray	ASTM A 48 class 20 to 60	(160 - 300) HB	$\sigma = 283\ HB - 22970$
Gray	ASTM A 48 class 20 to 60	(160 - 300) HB	$\sigma_c = 725\ HB - 29440$
Ductile	ASTM A, SAE, GGG, etc.	(140 - 390) HB	$\sigma = 450\ HB$
Ductile	ASTM A, SAE, GGG, etc.	(140 - 390) HB	$\sigma = 467\ HB - 4720$
Ductile	ASTM A 897-90 and GGG	(300 - 440) HB	$\sigma = 536\ HB - 7490$
Malleable	ASTM A: 220, 602, SAE J158	(150 - 270) HB	$\sigma = 400\ HB$
Malleable	ASTM A: 220, 602, SAE J158	(150 - 270) HB	$\sigma = 366\ HB + 6420$

Note: σ_c is compressive strength of gray cast iron

Table 9. Ultimate Strength – Hardness Relationship for Titanium Alloys and Pure Titanium

Work Material	Designation	Hardness range	Ultimate tensile strength
Titanium alloys	Alpha-beta alloys	(205 - 385) HB	$\sigma = 490\ HB$
	Alpha-beta alloys	(205 - 385) HB	$\sigma = 457\ HB + 9150$
	Alpha-beta alloys	(20 - 47) HRC	$\sigma = 3300\ HRC + 35230$
	Alpha-beta alloys	(230 - 420) HK	$\sigma = 450\ HK$
	Alpha-beta alloys	(230 - 420) HK	$\sigma = 440\ HK + 2910$
Pure titanium	ASTM grades 1 to 4	(120 - 215) HB	$\sigma = 510\ HB - 6160$
	ASTM grades 1 to 4	(140 - 240) HK	$\sigma = 478\ HK - 11730$

Table 10. Ultimate Strength – Hardness Relationship for Wrought Aluminum Alloys

Principal alloying elements	Designation	Hardness range	Ultimate tensile strength
Copper	2000 series	(45 - 130) HB	$\sigma = 580$ HB
Copper	2000 series	(45 - 130) HB	$\sigma = 555$ HB + 2150
Manganese	3000 series	(28 - 77) HB	$\sigma = 555$ HB
Manganese	3000 series	(28 - 77) HB	$\sigma = 522$ HB + 1595
Magnesium	5000 series, temper 0	(28 - 65) HB	$\sigma = 605$ HB
Magnesium	5000 series, temper 0	(28 - 65) HB	$\sigma = 617$ HB – 557
Magnesium	5000 series, temper H	(36 - 105) HB	$\sigma = 565$ HB
Magnesium	5000 series, temper H	(36 - 105) HB	$\sigma = 604$ HB – 2580
Magnesium and Silicon	6000 series, tempers: 0, T1, T4	(25 - 90) HB	$\sigma = 540$ HB
Magnesium and Silicon	6000 series, tempers: 0, T1, T4	(25 - 90) HB	$\sigma = 603$ HB – 3210
Magnesium and Silicon	6000 series, tempers: T5, T6, T83, T831, T832	(60 - 120) HB	$\sigma = 460$ HB
Magnesium and Silicon	6000 series, tempers: T5, T6, T83, T831, T832	(60 - 120) HB	$\sigma = 496$ HB – 2910
Zinc	7000 series	(60 - 150) HB	$\sigma = 554$ HB
Zinc	7000 series	(60 - 150) HB	$\sigma = 558$ HB – 409

Table 11. Ultimate Strength – Hardness Relationship for Cast Aluminum Alloys

Principal alloying elements	Designation	Hardness range	Ultimate tensile strength
Copper	204.0	(105 - 125) HB	$\sigma = 200$ HB + 36000
Copper	242.0	(70 - 110) HB	$\sigma = 446$ HB – 4465
Silicon and Copper	355.0	(65 - 105) HB	$\sigma = 448$ HB – 645
Silicon and Copper	356.0	(60 - 80) HB	$\sigma = 556$ HB – 6780
Silicon and Copper	390.0	(100 - 150) HB	$\sigma = 511$ HB – 24820
Zinc	771.0	(85 - 120) HB	$\sigma = 292$ HB + 13150

Table 12. Ultimate Strength – Hardness Relationship for Wrought Copper Alloys

Principal alloying elements	Designation	Hardness range	Ultimate tensile strength
Zinc	C2xxxx series	(55 - 82) HRB	σ = 933 HRB – 1990
Zinc	C2xxxx series	(100 - 160) HB	σ = 453 HB + 5150
Zinc and Lead	C3xxxx series	(60 - 85) HRB	σ = 796 HRB + 8370
Zinc and Lead	C3xxxx series	(110 - 165) HB	σ = 332 HB + 21900
Zinc and Tin	C46xxx series	(55 - 95) HRB	σ = 680 HRB + 19620
Zinc and Tin	C46xxx series	(100 - 210) HB	σ = 282 HB + 29890
Zinc, Tin, and Lead	C48xxx series	(55 - 82) HRB	σ = 546 HRB + 27820
Zinc, Tin, and Lead	C48xxx series	(100 - 160) HB	σ = 269 HB + 31350
Tin	C5xxxx series	(70 - 85) HRB	σ = 1259 HRB – 28520
Tin	C5xxxx series	(125 - 165) HB	σ = 474 HB + 810
Aluminum	C6xxxx series	(155 - 200) HB	σ = 583 HB – 4570
Silicon	C65xxx series	(60 - 95) HRB	σ = 1236 HRB – 19010
Silicon	C65xxx series	(110 - 210) HB	σ = 458 HB + 6670
Zinc, Aluminum, Nickel	C69000	(80 - 100) HRB	σ = 2137 HRB – 91620
Zinc, Aluminum, Nickel	C69000	(150 - 240) HB	σ = 506 HB + 3520

Table 13. Ultimate Strength – Hardness Relationship for Cast Copper Alloys

Principal alloying elements	Designation	Hardness range	Ultimate tensile strength
Zinc, Tin, and Lead	C8xxxx series	(50 - 80) HB	σ = 753 HB – 8220
Zinc and Aluminum	C86xxx series	(110 - 225) HB	σ = 472 HB + 11890
Zinc and Silicon	C87xxx series	(70 – 120) HB	σ = 265 HB + 35910
Tin, Lead, and Zinc	C92xxx, C93xxx series	(50 - 75) HB	σ = 761 HB – 10930
Nickel and Tin	C94-C96-C97xxx series	(60 - 175) HB	σ = 457 HB + 8220
Aluminum	C95xxx series	(140 - 250) HB	σ = 439 HB + 16860

6. Concluding Remarks

1. Metallic workpiece materials can be divided into two major groups: ferrous and nonferrous alloys.

Among **ferrous alloys**, *carbon steels* are by far the most frequently used workpiece materials (Ref. 5, p. 147). *Stainless steels* represent about 25% of the metals turned today. That percentage has been rising and will continue to rise with the growth of the chemical, oil, food processing, and power industries (Ref. 9, p. 24). Machining of stainless steels, especially the austenitic type, leads to more friction, heat, and chip-control problems.

Among **nonferrous metals**, *aluminum alloys* have found wide use in the automotive industry to fabricate engine blocks, cylinder heads, pistons, carburetors, wheels, etc. Aluminum alloys are machined at high-speed with superhard cutting tool materials such as diamond-film-coated inserts and polycrystalline diamond tips brazed onto carbide substrates. *Titanium alloys* are preferred metals for aerospace applications due to their high strength and high fatigue resistance. From a cutting viewpoint, titanium alloys are hard-to-machine work materials.

2. Hardness implies resistance to deformation, and in the case of metalcutting this characteristic is a measure of resistance to cutting. It is important to be able to convert one type of hardness measurement into another, especially if the work materials are characterized by different types of hardness. More than 35 linear regression equations were developed to express one type of hardness number in relation to another type.

3. Analytical study of the work materials' strength-hardness relationship produced more than 75 equations that can be used to calculate ultimate tensile and compressive strength of widely used work materials.

4. Knowledge of the mechanical properties of work materials leads to higher accuracy in calculation of cutting forces and machining power, and results in better guidelines for cutting tool applications. This knowledge defines the first element of the metalcutting as an integrated system: sufficient description of work materials.

5. Two other elements of an integrated metalcutting system – the cutting tool and the machine tool – are currently being studied by the author. This analytical and experimental study may result in a book discussing the dynamics of metalcutting.

Appendixes

Appendix I
Conversion of Hardness Numbers

1. Linear Regression Formulas for Hardness Conversion 29

2. Conversion of Rockwell B Hardness Numbers into Brinell Hardness Numbers 30

3. Conversion of Rockwell C Hardness Numbers into Brinell Hardness Numbers 37

4. Conversion of Vickers Hardness Numbers into Brinell Hardness Numbers 46

5. Conversion of Knoop Hardness Numbers into Brinell Hardness Numbers 53

6. Conversion of Scleroscope Hardness Numbers into Brinell Hardness Numbers 62

1. Linear Regression Formulas for Hardness Conversion

Statistical method of linear regression analysis was applied to the hardness conversion procedure by using a general equation of a straight line:

$$Y = aX \pm b \quad (1)$$

where ***X*** and ***Y*** are any two types of hardness numbers (for example, HRB and HB),

a is a slope, or the tangent of the angle made by a straight line with a horizontal coordinate axis,

b is an intercept, or the distance from the origin to a point where the line crosses a coordinate axis.

The slope is calculated by

$$a = \frac{\Sigma XYn - (\Sigma X)(\Sigma Y)}{\Sigma X^2 n - (\Sigma X)^2} \quad (2)$$

The intercept is obtained from

$$b = \frac{\Sigma Y - a\Sigma X}{n} \quad (3)$$

where ***n*** is the number of data points.

The strength of the linear relationship between any two types of hardness numbers is measured by the correlation coefficient r_{xy} that can be found from

$$r_{xy} = \frac{S_{xy}}{\sqrt{S_{xx} \cdot S_{yy}}} \quad (4)$$

where

$$S_{xy} = \Sigma XY - (\Sigma X)(\Sigma Y)/n$$

$$S_{xx} = \Sigma X^2 - (\Sigma X)^2/n$$

$$S_{yy} = \Sigma Y^2 - (\Sigma Y)^2/n$$

2. Conversion of Rockwell B Hardness Numbers into Brinell Hardness Numbers

BRINELL HARDNESS (500-kgf load) versus ROCKWELL HARDNESS, B-SCALE

$$HB = 0.897 \times HRB + 38.3$$

Rockwell hardness *HRB, X*	Brinell hardness *HB, Y*	*XY*	X^2	Y^2	Parameters of the regression line *slope*	Parameters of the regression line *intercept*	Correlation coefficient
41	75	3075	1681	5625	0.897	38.3	0.998
42	76	3192	1764	5776			
43	77	3311	1849	5929			
44	78	3432	1936	6084			
45	79	3555	2025	6241			
47	80	3760	2209	6400			
48	81	3888	2304	6561			
49	82	4018	2401	6724			
50	83	4150	2500	6889			
51	84	4284	2601	7056			
52	85	4420	2704	7225			
53	86	4558	2809	7396			
54	87	4698	2916	7569			
619	1053	50341	29699	85475			

BRINELL HARDNESS (3000-kgf load) versus ROCKWELL HARDNESS, B-SCALE

HB = 1.500 x HRB + 17.3

Rockwell hardness ***HRB, X***	Brinell hardness ***HB, Y***	***XY***	***X^2***	***Y^2***	Parameters of the regression line ***slope***	***intercept***	Correlation coefficient
55	100	5500	3025	10000	1.500	17.3	0.993
56	101	5656	3136	10201			
57	103	5871	3249	10609			
58	104	6032	3364	10816			
59	106	6254	3481	11236			
285	514	29313	16255	52862			

BRINELL HARDNESS (3000-kgf load) versus ROCKWELL HARDNESS, B-SCALE

HB = 1.800 x HRB - 1.4

Rockwell hardness	Brinell hardness	***XY***	X^2	Y^2	Parameters of the regression line		Correlation coefficient
HRB, X	***HB, Y***				***slope***	***intercept***	
60	107	6420	3600	11449	1.800	-1.4	0.999
61	108	6588	3721	11664			
62	110	6820	3844	12100			
63	112	7056	3969	12544			
64	114	7296	4096	12996			
65	116	7540	4225	13456			
66	117	7722	4356	13689			
67	119	7973	4489	14161			
68	121	8228	4624	14641			
69	123	8487	4761	15129			
645	1147	74130	41685	131829			

BRINELL HARDNESS (3000-kgf load) versus ROCKWELL HARDNESS, B-SCALE

HB = 2.394 x HRB - 42.7

Rockwell hardness	Brinell hardness	XY	X^2	Y^2	Parameters of the regression line		Correlation coefficient
HRB, X	***HB, Y***				***slope***	***intercept***	
70	125	8750	4900	15625	2.394	-42.7	0.999
71	127	9017	5041	16129			
72	130	9360	5184	16900			
73	132	9636	5329	17424			
74	135	9990	5476	18225			
75	137	10275	5625	18769			
76	139	10564	5776	19321			
77	141	10857	5929	19881			
78	144	11232	6084	20736			
79	147	11613	6241	21609			
745	1357	101294	55585	184619			

BRINELL HARDNESS (3000-kgf load) versus ROCKWELL HARDNESS, B-SCALE

HB = 3.297 x HRB - 114.4

Rockwell hardness ***HRB, X***	Brinell hardness ***HB, Y***	***XY***	***X^2***	***Y^2***	Parameters of the regression line		Correlation coefficient
					slope	***intercept***	
80	150	12000	6400	22500	3.297	-114.4	0.998
81	153	12393	6561	23409			
82	156	12792	6724	24336			
83	159	13197	6889	25281			
84	162	13608	7056	26244			
85	165	14025	7225	27225			
86	169	14534	7396	28561			
87	172	14964	7569	29584			
88	176	15488	7744	30976			
89	180	16020	7921	32400			
845	1642	139021	71485	270516			

BRINELL HARDNESS (3000-kgf load) versus ROCKWELL HARDNESS, B-SCALE

$$HB = 5.000 \times HRB - 265.0$$

Rockwell hardness *HRB, X*	Brinell hardness *HB, Y*	*XY*	X^2	Y^2	Parameters of the regression line: *slope*	Parameters of the regression line: *intercept*	Correlation coefficient
90	185	16650	8100	34225	5.000	-265.0	1.000
91	190	17290	8281	36100			
92	195	17940	8464	38025			
93	200	18600	8649	40000			
94	205	19270	8836	42025			
95	210	19950	9025	44100			
555	1185	109700	51355	234475			

BRINELL HARDNESS (3000-kgf load) versus ROCKWELL HARDNESS, B-SCALE

HB = 6.600 x HRB - 418.0

Rockwell hardness	Brinell hardness	*XY*	X^2	Y^2	Parameters of the regression line		Correlation coefficient
HRB, X	***HB, Y***				***slope***	***intercept***	
96	216	20736	9216	46656	6.600	-418.0	0.999
97	222	21534	9409	49284			
98	228	22344	9604	51984			
99	236	23364	9801	55696			
100	242	24200	10000	58564			
490	1144	112178	48030	262184			

3. Conversion of Rockwell C Hardness Numbers into Brinell Hardness Numbers

BRINELL HARDNESS (3000-kgf load) versus ROCKWELL HARDNESS, C-SCALE

$$HB = 5.329 \times HRC + 119.6$$

Rockwell hardness	Brinell hardness	***XY***	X^2	Y^2	Parameters of the regression line		Correlation coefficient
HRC, X	***HB, Y***				***slope***	***intercept***	
20.3	228	4628.4	412.09	51984	5.329	119.6	0.999
20.5	229	4694.5	420.25	52441			
21.0	231	4851.0	441.00	53361			
21.3	233	4962.9	453.69	54289			
21.7	235	5099.5	470.89	55225			
22.0	237	5214.0	484.00	56169			
22.2	238	5283.6	492.84	56644			
22.8	241	5494.8	519.84	58081			
23.0	243	5589.0	529.00	59049			
24.0	247	5928.0	576.00	61009			
24.2	248	6001.6	585.64	61504			
24.8	252	6249.6	615.04	63504			
25.0	253	6325.0	625.00	64009			
25.4	255	6477.0	645.16	65025			
318.2	3370	76798.9	7270.44	812294			

BRINELL HARDNESS (3000-kgf load) versus ROCKWELL HARDNESS, C-SCALE

HB = 6.984 x HRC + 76.1

Rockwell hardness	Brinell hardness	*XY*	X^2	Y^2	Parameters of the regression line		Correlation coefficient
HRC, X	***HB, Y***				***slope***	***intercept***	
26.0	258	6708.0	676.00	66564	6.984	76.1	0.999
26.4	261	6890.4	696.96	68121			
26.6	262	6969.2	707.56	68644			
27.0	264	7128.0	729.00	69696			
27.1	265	7181.5	734.41	70225			
27.6	269	7424.4	761.76	72361			
27.8	270	7506.0	772.84	72900			
28.0	271	7588.0	784.00	73441			
28.5	275	7837.5	812.25	75625			
28.8	277	7977.6	829.44	76729			
29.0	279	8091.0	841.00	77841			
29.2	280	8176.0	852.64	78400			
29.8	284	8463.2	888.04	80656			
29.9	285	8521.5	894.01	81225			
30.0	286	8580.0	900.00	81796			
421.7	4086	115042.3	11879.91	1114224			

BRINELL HARDNESS (3000-kgf load) versus ROCKWELL HARDNESS, C-SCALE

HB = 8.379 x HRC + 33.7

Rockwell hardness	Brinell hardness	XY	X^2	Y^2	Parameters of the regression line		Correlation coefficient
HRC, X	***HB, Y***				***slope***	***intercept***	
30.9	293	9053.7	954.81	85849	8.379	33.7	0.999
31.0	294	9114.0	961.00	86436			
32.0	301	9632.0	1024.00	90601			
32.1	302	9694.2	1030.41	91204			
32.2	303	9756.6	1036.84	91809			
33.0	311	10263.0	1089.00	96721			
33.3	313	10422.9	1108.89	97969			
34.0	319	10846.0	1156.00	101761			
34.3	321	11010.3	1176.49	103041			
34.4	322	11076.8	1183.36	103684			
35.0	327	11445.0	1225.00	106929			
35.5	331	11750.5	1260.25	109561			
397.7	3737	124065.0	13206.05	1165565			

BRINELL HARDNESS (3000-kgf load) versus ROCKWELL HARDNESS, C-SCALE

$$HB = 8.872 \times HRC + 16.0$$

Rockwell hardness *HRC, X*	Brinell hardness *HB, Y*	*XY*	X^2	Y^2	Parameters of the regression line *slope*	*intercept*	Correlation coefficient
35.5	331	11750.5	1260.25	109561	8.872	16.0	0.9997
36.0	336	12096.0	1296.00	112896			
36.6	341	12480.6	1339.56	116281			
37.0	344	12728.0	1369.00	118336			
37.7	350	13195.0	1421.29	122500			
37.9	352	13340.8	1436.41	123904			
38.0	353	13414.0	1444.00	124609			
38.8	360	13968.0	1505.44	129600			
39.0	362	14118.0	1521.00	131044			
39.1	363	14193.3	1528.81	131769			
39.8	369	14686.2	1584.04	136161			
40.0	371	14840.0	1600.00	137641			
40.4	375	15150.0	1632.16	140625			
495.8	4607	175960.4	18937.96	1634927			

BRINELL HARDNESS (3000-kgf load) versus ROCKWELL HARDNESS, C-SCALE

$$HB = 10.025 \times HRC - 30.8$$

Rockwell hardness *HRC, X*	Brinell hardness *HB, Y*	*XY*	X^2	Y^2	Parameters of the regression line *slope*	*intercept*	Correlation coefficient
40.8	379	15463.2	1664.64	143641	10.025	-30.8	0.998
41.0	381	15621.0	1681.00	145161			
41.8	388	16218.4	1747.24	150544			
42.0	390	16380.0	1764.00	152100			
42.7	397	16951.9	1823.29	157609			
43.0	400	17200.0	1849.00	160000			
43.1	401	17283.1	1857.61	160801			
43.6	405	17658.0	1900.96	164025			
44.0	409	17996.0	1936.00	167281			
44.5	415	18467.5	1980.25	172225			
45.0	421	18945.0	2025.00	177241			
45.3	425	19252.5	2052.09	180625			
516.8	4811	207436.6	22281.08	1931253			

BRINELL HARDNESS (3000-kgf load) versus ROCKWELL HARDNESS, C-SCALE

HB = 12.473 x HRC - 142.6

Rockwell hardness	Brinell hardness	*XY*	X^2	Y^2	Parameters of the regression line		Correlation coefficient
HRC, X	***HB, Y***				***slope***	***intercept***	
45.7	429	19605.3	2088.49	184041	12.473	-142.6	0.999
46.0	432	19872.0	2116.00	186624			
46.1	433	19961.3	2125.21	187489			
46.9	442	20729.8	2199.61	195364			
47.0	443	20821.0	2209.00	196249			
47.1	444	20912.4	2218.41	197136			
47.7	452	21560.4	2275.29	204304			
48.0	455	21840.0	2304.00	207025			
48.4	460	22264.0	2342.56	211600			
48.5	461	22358.5	2352.25	212521			
49.0	469	22981.0	2401.00	219961			
49.1	471	23126.1	2410.81	221841			
49.6	477	23659.2	2460.16	227529			
49.8	479	23854.2	2480.04	229441			
50.0	481	24050.0	2500.00	231361			
50.5	488	24644.0	2550.25	238144			
769.4	7316	352239.2	37033.08	3350630			

BRINELL HARDNESS (3000-kgf load) versus ROCKWELL HARDNESS, C-SCALE

HB = 15.962 x HRC - 318.7

Rockwell hardness *HRC, X*	Brinell hardness *HB, Y*	*XY*	X^2	Y^2	Parameters of the regression line *slope*	*intercept*	Correlation coefficient
51.0	495	25245.0	2601.00	245025	15.962	-318.7	0.999
51.0	496	25296.0	2601.00	246016			
51.1	497	25396.7	2611.21	247009			
51.7	507	26211.9	2672.89	257049			
52.0	512	26624.0	2704.00	262144			
52.1	514	26779.4	2714.41	264196			
52.3	517	27039.1	2735.29	267289			
53.0	525	27825.0	2809.00	275625			
53.5	534	28569.0	2862.25	285156			
53.6	535	28676.0	2872.96	286225			
54.0	543	29322.0	2916.00	294849			
54.1	545	29484.5	2926.81	297025			
54.7	554	30303.8	2992.09	306916			
54.7	555	30358.5	2992.09	308025			
55.0	560	30800.0	3025.00	313600			
55.2	564	31132.8	3047.04	318096			
849.0	8453	449063.7	45083.04	4474245			

BRINELL HARDNESS (3000-kgf load) versus ROCKWELL HARDNESS, C-SCALE

HB = 19.038 x HRC - 489.4

Rockwell hardness	Brinell hardness	*XY*	X^2	Y^2	Parameters of the regression line		Correlation coefficient
HRC, X	***HB, Y***				***slope***	***intercept***	
55.7	573	31916.1	3102.49	328329	19.038	-489.4	0.9996
56.0	577	32312.0	3136.00	332929			
56.3	582	32766.6	3169.69	338724			
56.8	591	33568.8	3226.24	349281			
57.0	595	33915.0	3249.00	354025			
57.3	601	34437.3	3283.29	361201			
57.8	611	35315.8	3340.84	373321			
58.0	615	35670.0	3364.00	378225			
58.3	620	36146.0	3398.89	384400			
58.7	627	36804.9	3445.69	393129			
58.8	630	37044.0	3457.44	396900			
59.0	634	37406.0	3481.00	401956			
59.2	638	37769.6	3504.64	407044			
59.7	647	38625.9	3564.09	418609			
60.0	653	39180.0	3600.00	426409			
60.1	656	39425.6	3612.01	430336			
928.7	9850	572303.6	53935.31	6074818			

BRINELL HARDNESS (3000-kgf load) versus ROCKWELL HARDNESS, C-SCALE

HB = 17.602 x HRC - 403.8

Rockwell hardness ***HRC, X***	Brinell hardness ***HB, Y***	***XY***	X^2	Y^2	Parameters of the regression line ***slope***	***intercept***	Correlation coefficient
61.0	670	40870.0	3721.00	448900	17.602	-403.8	0.999
61.7	682	42079.4	3806.89	465124			
61.8	684	42271.2	3819.24	467856			
62.0	688	42656.0	3844.00	473344			
62.5	698	43625.0	3906.25	487204			
63.0	705	44415.0	3969.00	497025			
63.3	710	44943.0	4006.89	504100			
63.4	712	45140.8	4019.56	506944			
64.0	722	46208.0	4096.00	521284			
64.7	733	47425.1	4186.09	537289			
65.0	739	48035.0	4225.00	546121			
65.3	745	48648.5	4264.09	555025			
65.9	757	49886.3	4342.81	573049			
66.4	767	50928.8	4408.96	588289			
890.0	10012	637132.1	56615.78	7171554			

4. Conversion of Vickers Hardness Numbers into Brinell Hardness Numbers

BRINELL HARDNESS (3000-kgf load) versus VICKERS HARDNESS

$$HB = 0.959 \times HV - 0.8$$

Vickers hardness *HV, X*	Brinell hardness *HB, Y*	*XY*	X^2	Y^2	Parameters of the regression line: *slope*	Parameters of the regression line: *intercept*	Correlation coefficient
85	81	6885	7225	6561	0.959	-0.8	0.9998
90	86	7740	8100	7396			
95	90	8550	9025	8100			
100	95	9500	10000	9025			
110	105	11550	12100	11025			
117	111	12987	13689	12321			
120	114	13680	14400	12996			
122	116	14152	14884	13456			
127	121	15367	16129	14641			
130	124	16120	16900	15376			
132	126	16632	17424	15876			
137	131	17947	18769	17161			
140	133	18620	19600	17689			
143	137	19591	20449	18769			
1648	1570	189321	198694	180392			

BRINELL HARDNESS (3000-kgf load) versus VICKERS HARDNESS

HB = 0.949 x HV + 0.9

Vickers hardness	Brinell hardness	*XY*	X^2	Y^2	Parameters of the regression line		Correlation coefficient
HV, X	***HB, Y***				***slope***	***intercept***	
150	143	21450	22500	20449	0.949	0.9	0.9998
153	146	22338	23409	21316			
156	149	23244	24336	22201			
159	152	24168	25281	23104			
163	156	25428	26569	24336			
167	159	26553	27889	25281			
170	162	27540	28900	26244			
171	163	27873	29241	26569			
175	167	29225	30625	27889			
178	170	30260	31684	28900			
182	174	31668	33124	30276			
188	179	33652	35344	32041			
190	181	34390	36100	32761			
192	183	35136	36864	33489			
196	187	36652	38416	34969			
2590	2471	429577	450282	409825			

BRINELL HARDNESS (3000-kgf load) versus VICKERS HARDNESS

HB = 0.954 x HV - 0.7

Vickers hardness	Brinell hardness	XY	X^2	Y^2	Parameters of the regression line		Correlation coefficient
HV, X	***HB, Y***				***slope***	***intercept***	
200	190	38000	40000	36100	0.954	-0.7	0.9997
202	192	38784	40804	36864			
207	197	40779	42849	38809			
210	200	42000	44100	40000			
212	201	42612	44944	40401			
218	207	45126	47524	42849			
220	209	45980	48400	43681			
222	212	47064	49284	44944			
228	217	49476	51984	47089			
230	219	50370	52900	47961			
234	223	52182	54756	49729			
240	228	54720	57600	51984			
241	229	55189	58081	52441			
245	233	57085	60025	54289			
247	235	58045	61009	55225			
3356	3192	717412	754260	682366			

BRINELL HARDNESS (3000-kgf load) versus VICKERS HARDNESS

$$HB = 0.922 \times HV + 7.3$$

Vickers hardness	Brinell hardness	*XY*	X^2	Y^2	Parameters of the regression line		Correlation coefficient
HV, X	***HB, Y***				***slope***	***intercept***	
250	238	59500	62500	56644	0.922	7.3	0.9997
253	241	60973	64009	58081			
255	243	61965	65025	59049			
260	247	64220	67600	61009			
261	248	64728	68121	61504			
265	252	66780	70225	63504			
269	255	68595	72361	65025			
270	256	69120	72900	65536			
275	261	71775	75625	68121			
276	262	72312	76176	68644			
280	265	74200	78400	70225			
284	269	76396	80656	72361			
285	270	76950	81225	72900			
290	275	79750	84100	75625			
292	277	80884	85264	76729			
295	280	82600	87025	78400			
4360	4139	1130748	1191212	1073357			

BRINELL HARDNESS (3000-kgf load) versus VICKERS HARDNESS

HB = 0.944 x HV + 1.2

Vickers hardness	Brinell hardness	*XY*	X^2	Y^2	Parameters of the regression line		Correlation coefficient
HV, X	***HB, Y***				***slope***	***intercept***	
300	284	85200	90000	80656	0.944	1.2	0.99996
310	294	91140	96100	86436			
320	303	96960	102400	91809			
330	313	103290	108900	97969			
340	322	109480	115600	103684			
350	331	115850	122500	109561			
360	341	122760	129600	116281			
370	350	129500	136900	122500			
380	360	136800	144400	129600			
390	369	143910	152100	136161			
396	375	148500	156816	140625			
3846	3642	1283390	1355316	1215282			

BRINELL HARDNESS (3000-kgf load) versus VICKERS HARDNESS

$$HB = 0.909 \times HV + 15.1$$

Vickers hardness *HV, X*	Brinell hardness *HB, Y*	*XY*	X^2	Y^2	Parameters of the regression line *slope*	*intercept*	Correlation coefficient
400	379	151600	160000	143641	0.909	15.1	0.9998
402	381	153162	161604	145161			
410	388	159080	168100	150544			
412	390	160680	169744	152100			
420	397	166740	176400	157609			
425	401	170425	180625	160801			
430	405	174150	184900	164025			
434	409	177506	188356	167281			
440	415	182600	193600	172225			
446	421	187766	198916	177241			
450	425	191250	202500	180625			
460	433	199180	211600	187489			
470	442	207740	220900	195364			
472	444	209568	222784	197136			
480	452	216960	230400	204304			
490	460	225400	240100	211600			
491	461	226351	241081	212521			
498	469	233562	248004	219961			
8030	7572	3393720	3599614	3199628			

BRINELL HARDNESS (3000-kgf load) versus VICKERS HARDNESS

HB = 0.940 x HV - 0.2

Vickers hardness *HV, X*	Brinell hardness *HB, Y*	*XY*	*X²*	*Y²*	Parameters of the regression line		Correlation coefficient
					slope	***intercept***	
500	471	235500	250000	221841	0.940	-0.2	0.9999
510	479	244290	260100	229441			
520	488	253760	270400	238144			
530	497	263410	280900	247009			
540	507	273780	291600	257049			
550	517	284350	302500	267289			
560	525	294000	313600	275625			
570	535	304950	324900	286225			
580	545	316100	336400	297025			
590	554	326860	348100	306916			
600	564	338400	360000	318096			
610	573	349530	372100	328329			
620	582	360840	384400	338724			
630	591	372330	396900	349281			
640	601	384640	409600	361201			
650	611	397150	422500	373321			
660	620	409200	435600	384400			
667	627	418209	444889	393129			
10527	9887	5827299	6204489	5473045			

5. Conversion of Knoop Hardness Numbers into Brinell Hardness Numbers

BRINELL HARDNESS (3000-kgf load) versus KNOOP HARDNESS

$$HB = 0.927 \times HK - 9.0$$

Knoop hardness *HK, X*	Brinell hardness *HB, Y*	*XY*	X^2	Y^2	Parameters of the regression line *slope*	*intercept*	Correlation coefficient
97	81	7857	9409	6561	0.927	-9.0	0.999
102	86	8772	10404	7396			
107	90	9630	11449	8100			
112	95	10640	12544	9025			
123	105	12915	15129	11025			
131	111	14541	17161	12321			
133	114	15162	17689	12996			
135	116	15660	18225	13456			
140	121	16940	19600	14641			
143	124	17732	20449	15376			
145	126	18270	21025	15876			
1368	1169	148119	173084	126773			

BRINELL HARDNESS (3000-kgf load) versus KNOOP HARDNESS

HB = 0.901 x HK - 4.6

Knoop hardness	Brinell hardness	*XY*	X^2	Y^2	Parameters of the regression line		Correlation coefficient
HK, X	*HB, Y*				*slope*	*intercept*	
151	131	19781	22801	17161	0.901	-4.6	0.999
154	133	20482	23716	17689			
157	137	21509	24649	18769			
163	143	23309	26569	20449			
164	143	23452	26896	20449			
166	146	24236	27556	21316			
170	149	25330	28900	22201			
174	152	26448	30276	23104			
175	152	26600	30625	23104			
178	156	27768	31684	24336			
182	159	28938	33124	25281			
185	162	29970	34225	26244			
186	163	30318	34596	26569			
190	167	31730	36100	27889			
194	170	32980	37636	28900			
196	171	33516	38416	29241			
198	174	34452	39204	30276			
2983	2608	460819	526973	402978			

BRINELL HARDNESS (3000-kgf load) versus KNOOP HARDNESS

$$HB = 0.994 \times HK - 23.5$$

Knoop hardness *HK, X*	Brinell hardness *HB, Y*	*XY*	X^2	Y^2	Parameters of the regression line *slope*	*intercept*	Correlation coefficient
202	179	36158	40804	32041	0.994	-23.5	0.998
206	181	37286	42436	32761			
207	183	37881	42849	33489			
212	187	39644	44944	34969			
216	190	41040	46656	36100			
217	192	41664	47089	36864			
222	197	43734	49284	38809			
226	200	45200	51076	40000			
227	201	45627	51529	40401			
232	207	48024	53824	42849			
234	209	48906	54756	43681			
237	212	50244	56169	44944			
242	217	52514	58564	47089			
243	219	53217	59049	47961			
247	223	55081	61009	49729			
3370	2997	676220	760038	601687			

BRINELL HARDNESS (3000-kgf load) versus KNOOP HARDNESS

HB = 0.952 x HK - 11.5

Knoop hardness	Brinell hardness	*XY*	X^2	Y^2	Parameters of the regression line		Correlation coefficient
HK, X	***HB, Y***				***slope***	***intercept***	
253	229	57937	64009	52441	0.952	-11.5	0.999
259	235	60865	67081	55225			
262	238	62356	68644	56644			
265	241	63865	70225	58081			
267	243	64881	71289	59049			
272	248	67456	73984	61504			
277	252	69804	76729	63504			
279	255	71145	77841	65025			
282	256	72192	79524	65536			
286	262	74932	81796	68644			
291	265	77115	84681	70225			
294	269	79086	86436	72361			
296	270	79920	87616	72900			
297	271	80487	88209	73441			
3880	3534	982041	1078064	894580			

BRINELL HARDNESS (3000-kgf load) versus KNOOP HARDNESS

$$HB = 1.030 \times HK - 34.3$$

Knoop hardness *HK, X*	Brinell hardness *HB, Y*	*XY*	X^2	Y^2	Parameters of the regression line		Correlation coefficient
					slope	*intercept*	
300	275	82500	90000	75625	1.030	-34.3	0.9995
302	277	83654	91204	76729			
304	279	84816	92416	77841			
305	280	85400	93025	78400			
309	284	87756	95481	80656			
310	285	88350	96100	81225			
311	286	88946	96721	81796			
318	293	93174	101124	85849			
326	301	98126	106276	90601			
327	302	98754	106929	91204			
328	303	99384	107584	91809			
334	311	103874	111556	96721			
336	311	104496	112896	96721			
337	313	105481	113569	97969			
342	319	109098	116964	101761			
345	321	110745	119025	103041			
346	322	111412	119716	103684			
5480	5062	1635966	1770586	1511632			

BRINELL HARDNESS (3000-kgf load) versus KNOOP HARDNESS

HB = 0.871 x HK + 21.5

Knoop hardness *HK, X*	Brinell hardness *HB, Y*	*XY*	X^2	Y^2	Parameters of the regression line		Correlation coefficient
					slope	*intercept*	
351	327	114777	123201	106929	0.871	21.5	0.999
356	331	117836	126736	109561			
360	336	120960	129600	112896			
367	341	125147	134689	116281			
370	344	127280	136900	118336			
378	350	132300	142884	122500			
379	352	133408	143641	123904			
380	353	134140	144400	124609			
389	360	140040	151321	129600			
391	362	141542	152881	131044			
392	363	142296	153664	131769			
4113	3819	1429726	1539917	1327429			

BRINELL HARDNESS (3000-kgf load) versus KNOOP HARDNESS

HB = 0.774 x HK + 60.0

Knoop hardness *HK, X*	Brinell hardness *HB, Y*	*XY*	X^2	Y^2	Parameters of the regression line		Correlation coefficient
					slope	*intercept*	
400	369	147600	160000	136161	0.774	60.0	0.9998
402	371	149142	161604	137641			
407	375	152625	165649	140625			
412	379	156148	169744	143641			
414	381	157734	171396	145161			
423	388	164124	178929	150544			
426	390	166140	181476	152100			
435	397	172695	189225	157609			
438	400	175200	191844	160000			
441	401	176841	194481	160801			
447	405	181035	199809	164025			
452	409	184868	204304	167281			
459	415	190485	210681	172225			
466	421	196186	217156	177241			
471	425	200175	221841	180625			
476	429	204204	226576	184041			
480	432	207360	230400	186624			
482	433	208706	232324	187489			
494	442	218348	244036	195364			
495	443	219285	245025	196249			
496	444	220224	246016	197136			
9416	8549	3849125	4242516	3492583			

BRINELL HARDNESS (3000-kgf load) versus KNOOP HARDNESS

HB = 0.841 x HK + 26.2

Knoop hardness ***HK, X***	Brinell hardness ***HB, Y***	***XY***	X^2	Y^2	Parameters of the regression line		Correlation coefficient
					slope	***intercept***	
505	452	228260	255025	204304	0.841	26.2	0.9995
517	460	237820	267289	211600			
518	461	238798	268324	212521			
528	471	248688	278784	221841			
537	477	256149	288369	227529			
539	479	258181	290521	229441			
550	488	268400	302500	238144			
558	495	276210	311364	245025			
561	497	278817	314721	247009			
572	507	290004	327184	257049			
579	514	297606	335241	264196			
583	517	301411	339889	267289			
594	525	311850	352836	275625			
7141	6343	3492194	3932047	3101573			

BRINELL HARDNESS (3000-kgf load) versus KNOOP HARDNESS

HB = 0.919 x HK - 20.2

Knoop hardness *HK, X*	Brinell hardness *HB, Y*	*XY*	X^2	Y^2	Parameters of the regression line *slope*	*intercept*	Correlation coefficient
604	534	322536	364816	285156	0.919	-20.2	0.9996
604	535	323140	364816	286225			
612	543	332316	374544	294849			
615	545	335175	378225	297025			
625	554	346250	390625	306916			
626	555	347430	391876	308025			
630	560	352800	396900	313600			
636	564	358704	404496	318096			
646	573	370158	417316	328329			
650	577	375050	422500	332929			
652	578	376856	425104	334084			
657	582	382374	431649	338724			
667	591	394197	444889	349281			
670	595	398650	448900	354025			
677	601	406877	458329	361201			
687	611	419757	471969	373321			
690	615	424350	476100	378225			
697	620	432140	485809	384400			
703	627	440781	494209	393129			
12348	10960	7139541	8043072	6337540			

6. Conversion of Scleroscope Hardness Numbers into Brinell Hardness Numbers

BRINELL HARDNESS (3000-kgf load) versus SCLEROSCOPE HARDNESS

$$HB = 7.000 \times HS - 11.5$$

Scleroscope hardness	Brinell hardness	*XY*	X^2	Y^2	Parameters of the regression line		Correlation coefficient
HS, X	***HB, Y***				***slope***	***intercept***	
17	109	1853	289	11881	7.000	-11.5	0.999
18	114	2052	324	12996			
19	121	2299	361	14641			
20	127	2540	400	16129			
21	135	2835	441	18225			
22	143	3146	484	20449			
23	150	3450	529	22500			
24	157	3768	576	24649			
25	163	4075	625	26569			
26	170	4420	676	28900			
27	179	4833	729	32041			
28	183	5124	784	33489			
29	192	5568	841	36864			
299	1943	45963	7059	299333			

BRINELL HARDNESS (3000-kgf load) versus SCLEROSCOPE HARDNESS

$$HB = 7.006 \times HS - 11.5$$

Scleroscope hardness	Brinell hardness	*XY*	X^2	Y^2	Parameters of the regression line		Correlation coefficient
HS, X	***HB, Y***				***slope***	***intercept***	
30	199	5970	900	39601	7.006	-11.5	0.9995
31	206	6386	961	42436			
32	212	6784	1024	44944			
33	219	7227	1089	47961			
34	228	7752	1156	51984			
35	233	8155	1225	54289			
36	241	8676	1296	58081			
37	247	9139	1369	61009			
38	255	9690	1444	65025			
39	262	10218	1521	68644			
345	2302	79997	11985	533974			

BRINELL HARDNESS (3000-kgf load) versus SCLEROSCOPE HARDNESS

$$HB = 7.428 \times HS - 27.6$$

Scleroscope hardness	Brinell hardness	XY	X^2	Y^2	Parameters of the regression line		Correlation coefficient
HS, X	***HB, Y***				***slope***	***intercept***	
40	268	10720	1600	71824	7.428	-27.6	0.999
41	277	11357	1681	76729			
42	285	11970	1764	81225			
43	293	12599	1849	85849			
44	301	13244	1936	90601			
46	312	14352	2116	97344			
47	321	15087	2209	103041			
48	330	15840	2304	108900			
49	336	16464	2401	112896			
400	2723	121633	17860	828409			

BRINELL HARDNESS (3000-kgf load) versus SCLEROSCOPE HARDNESS

$$HB = 7.771 \times HS - 45.3$$

Scleroscope hardness	Brinell hardness	XY	X^2	Y^2	Parameters of the regression line		Correlation coefficient
HS, X	***HB, Y***				***slope***	***intercept***	
50	342	17100	2500	116964	7.771	-45.3	0.998
51	352	17952	2601	123904			
52	362	18824	2704	131044			
54	373	20142	2916	139129			
55	380	20900	3025	144400			
56	389	21784	3136	151321			
57	398	22686	3249	158404			
58	405	23490	3364	164025			
59	415	24485	3481	172225			
492	3416	187363	26976	1301416			

BRINELL HARDNESS (3000-kgf load) versus SCLEROSCOPE HARDNESS

HB = 9.750 x HS - 167.3

Scleroscope hardness	Brinell hardness	*XY*	X^2	Y^2	Parameters of the regression line		Correlation coefficient
HS, X	***HB, Y***				***slope***	***intercept***	
60	421	25260	3600	177241	9.750	-167.3	0.996
61	429	26169	3721	184041			
63	444	27972	3969	197136			
64	455	29120	4096	207025			
65	461	29965	4225	212521			
66	477	31482	4356	227529			
67	488	32696	4489	238144			
68	497	33796	4624	247009			
69	507	34983	4761	257049			
583	4179	271443	37841	1947695			

BRINELL HARDNESS (3000-kgf load) versus SCLEROSCOPE HARDNESS

HB = 12.321 x HS - 346.4

Scleroscope hardness	Brinell hardness	*XY*	X^2	Y^2	Parameters of the regression line		Correlation coefficient
HS, X	***HB, Y***				***slope***	***intercept***	
70	516	36120	4900	266256	12.321	-346.4	0.999
71	525	37275	5041	275625			
72	545	39240	5184	297025			
73	554	40442	5329	306916			
74	564	41736	5476	318096			
75	577	43275	5625	332929			
76	591	44916	5776	349281			
77	601	46277	5929	361201			
78	615	47970	6084	378225			
79	627	49533	6241	393129			
745	5715	426784	55585	3278683			

Appendix II

Statistical Treatment of Data

Introduction

Any testing procedure must include the collection of experimental data, and its analysis and interpretation by statistical treatment. Unfortunately, in many cases our tests are limited to the collection of data and some analysis. Therefore, we cannot conclude whether or not the test results are sufficient, unless the statistical treatment of data is performed. Statistical treatment is a necessary scientific tool to make decision concerning the data points of a population in respect to data points contained in a sample. The following parameters should be calculated to obtain the necessary information about the sample:

1. Sample mean **X**
2. Sample standard deviation **S**
3. Coefficient of variation **V**
4. Absolute error of the sample mean α
5. Relative error of the sample mean ε
6. Confidence interval for the population mean $\mathbf{X - \alpha t \leq \mu \leq X + \alpha t}$

where μ is a population mean,
t is a value of Student's distribution versus probability of error at various degrees of freedom (d.f.). Values of Student's **t** distribution are shown in the table, p. 73.

1. Sample Mean

Sample mean is a value about which the data is centered. Such value is called a ***measure of average***. There are several types of such measures, the most common of which is the arithmetic mean. The sample mean is calculated as:

$$X = \frac{\sum_{i=1}^{n} X_i}{n} \qquad (1)$$

where X_i is an individual data point
n is the number of data points.

2. Sample Standard Deviation

Sample standard deviation is a measure of the dispersion of data about its mean. The sample standard deviation is calculated by the formula:

$$S = \sqrt{\frac{\sum_{i=1}^{n} (X_i - X)^2}{n - 1}} \qquad (2)$$

Where n-1 is the number of degrees of freedom (d.f.).

Sample standard deviation has the same unit of measure as the sample mean.

3. Coefficient of Variation

The coefficient of variation is used to evaluate or control the variability in data points. The coefficient of variation is calculated by dividing the sample standard deviation by the sample mean and expressing the result in percent:

$$V = \frac{S}{X} 100\% \qquad (3)$$

This quantity is also known as relative variability.

4. Absolute Error of the Sample Mean

The absolute error of the sample mean is calculated by dividing the sample standard deviation by the square root of the number of data points:

$$\alpha = \frac{S}{\sqrt{n}} \qquad (4)$$

Absolute error of the sample mean has the same unit of measure as the sample standard deviation and the sample mean.

5. Relative Error of the Sample Mean

The relative error of the sample mean is obtained by dividing the absolute error of the sample mean by the sample mean and expressing the result in percent:

$$\varepsilon = \frac{\alpha}{X} 100\% \qquad (5)$$

6. Confidence Interval for the Population Mean

Formulas (1) through (5) are based on a sample mean, which is only an estimate of the true (population) mean. Therefore, it is important to define a confidence interval that provides a range within which the true (population) mean lies at a given degree of confidence. A confidence interval for the population mean is obtained from:

$$X - \alpha t \leq \mu \leq X + \alpha t \qquad (6)$$

where X is a sample mean
α is an absolute error of the sample mean
t is a value of Student's distribution defined by the probability of error and by the number of degrees of freedom.

A 95% two-sided confidence interval is recommended for statistical analysis. In this case, the probability of error is ± 5% or ± 2.5% per side. The table for critical values of ***t*** distribution (page 73) should be used to locate the ***t*** value in the column under the heading "$t_{.025}$" with respect to given degrees of freedom (d.f.). A 5% probability of error provides practical accuracy, which is commonly acceptable in various engineering calculations.

Critical Values of *t* Distribution

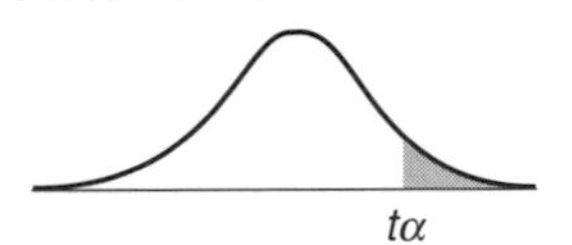

d.f.	$t_{.100}$	$t_{.050}$	$t_{.025}$	$t_{.010}$	$t_{.005}$	d.f.
1	3.078	6.314	12.706	31.821	63.657	1
2	1.886	2.920	4.303	6.965	9.925	2
3	1.638	2.353	3.182	4.541	5.841	3
4	1.533	2.132	2.776	3.747	4.604	4
5	1.476	2.015	2.571	3.365	4.032	5
6	1.440	1.943	2.447	3.143	3.707	6
7	1.415	1.895	2.365	2.998	3.499	7
8	1.397	1.860	2.306	2.896	3.355	8
9	1.383	1.833	2.262	2.821	3.250	9
10	1.372	1.812	2.228	2.764	3.169	10
11	1.363	1.796	2.201	2.718	3.106	11
12	1.356	1.782	2.179	2.681	3.055	12
13	1.350	1.771	2.160	2.650	3.012	13
14	1.345	1.761	2.145	2.624	2.977	14
15	1.341	1.753	2.131	2.602	2.947	15
16	1.337	1.746	2.120	2.583	2.921	16
17	1.333	1.740	2.110	2.567	2.898	17
18	1.330	1.734	2.101	2.552	2.878	18
19	1.328	1.729	2.093	2.539	2.861	19
20	1.325	1.725	2.086	2.528	2.845	20
21	1.323	1.721	2.080	2.518	2.831	21
22	1.321	1.717	2.074	2.508	2.819	22
23	1.319	1.714	2.069	2.500	2.807	23
24	1.318	1.711	2.064	2.492	2.797	24
25	1.316	1.708	2.060	2.485	2.787	25
26	1.315	1.706	2.056	2.479	2.779	26
27	1.314	1.703	2.052	2.473	2.771	27
28	1.313	1.701	2.048	2.467	2.763	28
29	1.311	1.699	2.045	2.462	2.756	29
inf.	1.282	1.645	1.960	2.326	2.576	inf.

Source: From "Table of Percentage Points of the t-Distribution," *Biometrika 32* (1941) 300. Reproduced by permission of the *Biometrika* Trustees.

Appendix III

Hardness and Tensile Strength Relationship for Carbon Steels

Relationship Between Hardness and Tensile Strength

AISI 1015 Low-Carbon Steel, Hardness range (100-220) HB

Brinell hardness, HB (3000-kgf load)	Tensile strength, psi σ	Ratio σ/HB
101	50000	495
111	56000	505
111	56000	505
111	56000	505
116	59250	511
116	60000	517
121	61500	508
121	62000	512
126	61000	484
126	63250	502
156	75500	484
217	106250	490
1533	**766750**	**6018**
average: 128	**average: 63896**	**average: 500**

Statistical Analysis of Data for the Tensile Strength (σ)-Brinell Hardness (HB) Ratio (σ/HB)

AISI 1015 Low-Carbon Steel, Hardness range (100 - 220) HB

Observation	σ/HB	σ/HB-X	$(\sigma/HB-X)^2$
1	495	-5	25
2	505	5	25
3	505	5	25
4	505	5	25
5	511	11	121
6	517	17	289
7	508	8	64
8	512	12	144
9	484	-16	256
10	502	2	4
11	484	-16	256
12	490	-10	100
Average ratio: **X** = 500		**Sum of squares:** $\Sigma(\sigma/HB-X)^2 = 1334$	

Nomenclature	Symbol	Value
1. Number of observations	**n**	12
2. Number of degrees of freedom (d.f.)	**n - 1**	11
3. Sample mean (average σ/HB ratio)	**X**	500
4. Absolute error of the mean	α	3.2
5. Average σ/HB ratio range	$X \pm \alpha$	500 ± 3
6. Sample standard deviation	**S**	11.0
7. Coefficient of variation	**V**	2.2%
8. Relative error of the mean	ε	0.6%
9. *t* value at 95% two-sided confidence interval for d.f. = 11	$t_{.025}$	2.201
10. Population mean is greater than:	$X-\alpha_x t$	493
11. Population mean is less than:	$X+\alpha_x t$	507

Relationship Between Hardness and Tensile Strength

Low-Carbon Steels, Hardness range (100-160) HB

AISI grade	Brinell hardness, HB (3000-kgf load)	Tensile strength, psi σ	Ratio σ/HB
1016	111	55000	495
1016	121	61000	504
1017	105	53000	505
1017	116	59000	509
1018	101	50000	495
1018	111	55000	495
1018	111	55000	495
1018	116	58000	500
1018	121	60000	496
1018	121	60000	496
1018	126	64000	508
1018	131	65000	496
1018	131	65000	496
1018	143	69000	483
1018	143	70000	490
1018	163	82000	503
1019	116	59000	509
1019	131	66000	504
	2218	**1106000**	**8979**
	average: 123	**average: 61444**	**average: 499**

Relationship Between Hardness and Tensile Strength

AISI 1020 Low-Carbon Steel, Hardness range (110-255) HB

Brinell hardness, HB (3000-kgf load)	Tensile strength, psi σ	Ratio σ/HB
111	55000	495
111	57250	516
121	60000	496
121	61000	504
126	63500	504
131	64000	489
131	64000	489
131	64500	492
137	68500	500
143	71250	498
156	75500	484
179	87000	486
255	129000	506
1853	**920500**	**6459**
average: 143	**average: 70808**	**average: 497**

Statistical Analysis of Data for the Tensile Strength (σ)-Brinell Hardness (HB) Ratio (σ/HB)

AISI 1020 Low-Carbon Steel, Hardness range (110 - 255) HB

Observation	σ/HB	σ/HB-X	$(\sigma/HB-X)^2$
1	495	-5	25
2	516	16	256
3	496	-4	16
4	504	4	16
5	504	4	16
6	489	-11	121
7	489	-11	121
8	492	-8	64
9	500	0	0
10	498	-2	4
11	484	-16	256
12	486	-14	196
13	506	6	36
Average ratio: **X** = 500		**Sum of squares:** $\Sigma(\sigma/HB-X)^2 = 1127$	

Nomenclature	Symbol	Value
1. Number of observations	**n**	13
2. Number of degrees of freedom (d.f.)	**n - 1**	12
3. Sample mean (average σ/HB ratio)	**X**	500
4. Absolute error of the mean	α	2.7
5. Average σ/HB ratio range	$X \pm \alpha$	500 ± 3
6. Sample standard deviation	**S**	9.7
7. Coefficient of variation	**V**	1.9%
8. Relative error of the mean	ε	0.5%
9. *t* value at 95% two-sided confidence interval for d.f. = 12	$t_{.025}$	2.179
10. Population mean is greater than:	$X - \alpha_x t$	494
11. Population mean is less than:	$X + \alpha_x t$	506

Relationship Between Hardness and Tensile Strength

Low-Carbon Steels, Hardness range (100-140) HB

AISI grade	Brinell hardness, HB (3000-kgf load)	Tensile strength, psi σ	Ratio σ/HB
1023	111	56000	505
1023	121	62000	512
1025	101	50000	495
1025	111	55000	495
1025	111	55000	495
1025	116	58000	500
1025	121	60000	496
1025	121	60000	496
1025	126	64000	508
1025	131	65000	496
1025	131	65000	496
1025	143	70000	490
1026	126	64000	508
1026	143	71000	497
	1713	**855000**	**6989**
	average: 122	**average: 61071**	**average: 499**

Relationship Between Hardness and Tensile Strength

Low-Carbon Free-machining Steels, Hardness range (110-170) HB

AISI grade	Brinell hardness, HB (3000-kgf load)	Tensile strength, psi σ	Ratio σ/HB
1117	121	62000	512
1117	121	62000	512
1117	137	68000	496
1117	137	69000	504
1117	143	71000	497
1117	170	86000	506
1118	111	55000	495
1118	116	58000	500
1118	121	60000	496
1118	121	60000	496
1118	126	64000	508
1118	131	65000	496
1118	131	65000	496
1118	131	65000	496
1118	131	65000	496
1118	143	69000	483
1118	143	70000	490
1118	143	70000	490
1118	143	70000	490
1118	149	75000	503
1118	149	75000	503
1118	149	76000	510
1118	163	80000	491
	3130	**1560000**	**11466**
	average: 136	**average: 67826**	**average: 498**

Relationship Between Hardness and Tensile Strength

AISI 1035 Medium-Carbon Steel, Hardness range (130-210) HB

Brinell hardness, HB (3000-kgf load)	Tensile strength, psi σ	Ratio σ/HB
131	65000	496
143	70000	490
143	70000	490
143	72000	503
149	75000	503
149	75000	503
149	75000	503
163	80000	491
163	80000	491
163	80000	491
163	80000	491
170	85000	500
170	85000	500
179	90000	503
207	103000	498
2305	**1185000**	**7453**
average: 159	**average: 79000**	**average: 497**

Statistical Analysis of Data for the Tensile Strength (σ)-Brinell Hardness (HB) Ratio (σ/HB)

AISI 1035 Carbon Steel

Observation	σ/HB	σ/HB-X	$(\sigma/HB-X)^2$
1	496	-4	16
2	490	-10	100
3	490	-10	100
4	503	3	9
5	503	3	9
6	503	3	9
7	503	3	9
8	491	-9	81
9	491	-9	81
10	491	-9	81
11	491	-9	81
12	500	0	0
13	500	0	0
14	503	3	9
15	498	-2	4
Average ratio: X = 500		**Sum of squares:** $\Sigma(\sigma/HB-X)^2 = 589$	

Nomenclature	Symbol	Value
1. Number of observations	**n**	15
2. Number of degrees of freedom (d.f.)	**n - 1**	14
3. Sample mean (average σ/HB ratio)	**X**	500
4. Absolute error of the mean	α	1.7
5. Average σ/HB ratio range	**X ± α**	500 ± 2
6. Sample standard deviation	**S**	6.5
7. Coefficient of variation	**V**	1.3%
8. Relative error of the mean	ε	0.3%
9. *t* value at 95% two-sided confidence interval for d.f. = 14	$t_{.025}$	2.145
10. Population mean is greater than:	$X-\alpha_x t$	496
11. Population mean is less than:	$X+\alpha_x t$	504

Relationship Between Hardness and Tensile Strength

Medium-Carbon Steels, Hardness range (140-190) HB

AISI grade	Brinell hardness, HB (3000-kgf load)	Tensile strength, psi σ	Ratio σ/HB
1038	149	75000	503
1038	163	83000	509
1039	156	79000	506
1039	179	88000	492
1040	143	70000	490
1040	149	75000	503
1040	149	75000	503
1040	149	75000	503
1040	149	76000	510
1040	163	80000	491
1040	163	80000	491
1040	163	80000	491
1040	170	85000	500
1040	170	85000	500
1040	170	85000	500
1040	170	85000	500
1040	170	86000	506
1040	178	90000	506
1040	187	95000	508
	3090	**1547000**	**9512**
	average: 163	**average: 81421**	**average: 501**

Relationship Between Hardness and Tensile Strength

Medium-Carbon Steels, Hardness range (160-200) HB

AISI grade	Brinell hardness, HB (3000-kgf load)	Tensile strength, psi σ	Ratio σ/HB
1044	163	80000	491
1045	163	80000	491
1045	163	82000	503
1045	170	85000	500
1045	170	85000	500
1045	170	85000	500
1045	170	85000	500
1045	179	90000	503
1045	179	90000	503
1045	179	90000	503
1045	179	91000	508
1045	187	95000	508
1045	187	95000	508
1045	187	95000	508
1045	197	100000	508
1046	170	85000	500
1046	179	90000	503
1046	187	94000	503
	3179	**1597000**	**9040**
	average: 177	**average: 88722**	**average: 502**

Relationship Between Hardness and Tensile Strength

AISI 1050 Medium-Carbon Steel, Hardness range (170-270) HB

Brinell hardness, HB (3000-kgf load)	Tensile strength, psi σ	Ratio σ/HB
170	85000	500
170	85000	500
179	90000	503
179	90000	503
179	90000	503
187	92200	493
187	95000	508
187	95000	508
187	95000	508
189	95000	503
197	96700	491
197	100000	508
197	100000	508
197	100000	508
201	100000	498
207	101000	488
212	106200	501
217	104500	482
217	105000	484
217	106000	488
217	108500	500
217	109000	502
223	107700	483
223	111500	500
223	112000	502
223	114000	511
229	110000	480
229	112200	490
229	112500	491
235	117200	499
241	118000	490
241	119000	494
241	121000	502
248	122000	492
248	122500	494
248	122700	495
248	123500	498
255	129500	508
262	131200	501
262	132500	506
269	134000	498
8884	**4422100**	**20421**
average: 217	**average: 107856**	**average: 498**

Relationship Between Hardness and Tensile Strength

AISI 1141 Medium-Carbon Free-machining Steel, Hardness range (170-260) HB

Brinell hardness, HB (3000-kgf load)	Tensile strength, psi σ	Ratio σ/HB
170	85000	500
179	90000	503
183	95000	519
187	95000	508
192	96000	500
192	98000	510
197	96000	487
197	97000	492
197	100000	508
201	100000	498
201	101000	502
201	101000	502
201	102000	507
201	103000	512
207	103000	498
207	106000	512
212	105000	495
212	107000	505
217	105000	484
217	108000	498
223	110000	493
223	112000	502
229	110000	480
235	116000	494
262	130000	496
5143	**2571000**	**12505**
average: 206	**average: 102840**	**average: 500**

Relationship Between Hardness and Tensile Strength

Medium-Carbon Manganese Steels, Hardness range (190-220) HB

AISI grade	Brinell hardness, HB (3000-kgf load)	Tensile strength, psi σ	Ratio σ/HB
1547	187	95000	508
1547	192	94000	490
1547	207	103000	498
1548	192	94000	490
1548	197	96000	487
1548	217	106000	488
	1192	**588000**	**2961**
	average: 199	**average: 98000**	**average: 493**

Relationship Between Hardness and Tensile Strength

AISI 1060 High-Carbon Steel, Hardness range (180-300) HB

Brinell hardness, HB (3000-kgf load)	Tensile strength, psi σ	Ratio σ/HB
179	91000	508
179	91000	508
183	90000	492
201	96000	478
223	108000	484
223	110000	493
229	112000	489
229	113000	493
229	113000	493
241	118000	490
241	119000	494
248	124000	500
248	125000	504
255	128000	502
262	132000	504
262	133000	508
269	135000	502
269	136000	506
277	140000	505
285	143000	502
293	146000	498
302	149000	493
5327	**2652000**	**10946**
average: 242	**average: 120545**	**average: 498**

Statistical Analysis of Data for the Tensile Strength (σ)-Brinell Hardness (HB) Ratio (σ/HB)

AISI 1060 High-Carbon Steel, Hardness range (180-300) HB

Observation	σ/HB	σ/HB-X	$(\sigma/HB-X)^2$
1	508	8	64
2	508	8	64
3	492	-8	64
4	478	-22	484
5	484	-16	256
6	493	-7	49
7	489	-11	121
8	493	-7	49
9	493	-7	49
10	490	-10	100
11	494	-6	36
12	500	0	0
13	504	4	16
14	502	2	4
15	504	4	16
16	508	8	64
17	502	2	4
18	506	6	36
19	505	5	25
20	502	2	4
21	498	-2	4
22	493	-7	49
Average ratio: X = 500		**Sum of squares:** $\Sigma(\sigma/HB-X)^2 = 1558$	

Nomenclature	Symbol	Value
1. Number of observations	n	22
2. Number of degrees of freedom (d.f.)	n - 1	21
3. Sample mean (average σ/HB ratio)	X	500
4. Absolute error of the mean	α	1.8
5. Average σ/HB ratio range	X ± α	500 ± 2
6. Sample standard deviation	S	8.6
7. Coefficient of variation	V	1.7%
8. Relative error of the mean	ε	0.4%
9. *t* value at 95% two-sided confidence interval for d.f. = 21	$t_{.025}$	2.080
10. Population mean is greater than:	$X-\alpha_x t$	496
11. Population mean is less than:	$X+\alpha_x t$	504

Relationship Between Hardness and Tensile Strength

AISI 1080 High-Carbon Steel, Hardness range (170-360) HB

Brinell hardness, HB (3000-kgf load)	Tensile strength, psi σ	Ratio σ/HB
174	89000	511
174	89000	511
192	98000	510
229	112000	489
269	134000	498
269	135000	502
277	140000	505
285	141000	495
293	140000	478
293	146000	498
293	150000	512
302	150000	497
302	152000	503
311	157000	505
321	164000	511
331	166000	502
341	169000	496
341	171000	501
352	180000	511
352	182000	517
363	184000	507
6064	**3049000**	**10559**
average: 289	**average: 145190**	**average: 503**

Statistical Analysis of Data for the Tensile Strength (σ)-Brinell Hardness (HB) Ratio (σ/HB)

AISI 1080 High-Carbon Steel, Hardness range (170-360) HB

Observation	σ/HB	σ/HB-X	$(\sigma/HB-X)^2$
1	511	11	121
2	511	11	121
3	510	10	100
4	489	-11	121
5	498	-2	4
6	502	2	4
7	505	5	25
8	495	-5	25
9	478	-22	484
10	498	-2	4
11	512	12	144
12	497	-3	9
13	503	3	9
14	505	5	25
15	511	11	121
16	502	2	4
17	496	-4	16
18	501	1	1
19	511	11	121
20	517	17	289
21	507	7	49
Average ratio: **X** = 500		**Sum of squares:** $\Sigma(\sigma/HB-X)^2$ = 1797	

Nomenclature	Symbol	Value
1. Number of observations	**n**	21
2. Number of degrees of freedom (d.f.)	**n - 1**	20
3. Sample mean (average σ/HB ratio)	**X**	500
4. Absolute error of the mean	α	2.1
5. Average σ/HB ratio range	**X** $\pm \alpha$	500 ± 2
6. Sample standard deviation	**S**	9.5
7. Coefficient of variation	**V**	1.9%
8. Relative error of the mean	ε	0.4%
9. *t* value at 95% two-sided confidence interval for d.f. = 20	$t_{.025}$	2.086
10. Population mean is greater than:	$X-\alpha_x t$	496
11. Population mean is less than:	$X+\alpha_x t$	504

Relationship Between Hardness and Tensile Strength

AISI 1095 High-Carbon Steel, Hardness range (190-360) HB

Brinell hardness, HB (3000-kgf load)	Tensile strength, psi σ	Ratio σ/HB
192	95000	495
192	95000	495
197	99000	503
248	120000	484
255	128000	502
262	130000	496
269	132000	491
269	134000	498
277	140000	505
293	140000	478
293	142000	485
293	147000	502
293	147000	502
302	148000	490
302	151000	500
311	151000	486
321	160000	498
331	165000	498
331	166000	502
331	168000	508
352	176000	500
363	184000	507
6277	**3118000**	**10925**
average: 285	**average: 141727**	**average: 497**

Statistical Analysis of Data for the Tensile Strength (σ)-Brinell Hardness (HB) Ratio (σ/HB)

AISI 1095 High-Carbon Steel, Hardness range (190-360) HB

Observation	σ/HB	σ/HB-X	$(\sigma/HB-X)^2$
1	495	-5	25
2	495	-5	25
3	503	3	9
4	484	-16	256
5	502	2	4
6	496	-4	16
7	491	-9	81
8	498	-2	4
9	505	5	25
10	478	-22	484
11	485	-15	225
12	502	2	4
13	502	2	4
14	490	-10	100
15	500	0	0
16	486	-14	196
17	498	-2	4
18	498	-2	4
19	502	2	4
20	508	8	64
21	500	0	0
22	507	7	49
Average ratio: X = 500		**Sum of squares:** $\Sigma(\sigma/HB-X)^2$ = 1583	

Nomenclature	Symbol	Value
1. Number of observations	n	22
2. Number of degrees of freedom (d.f.)	n - 1	21
3. Sample mean (average σ/HB ratio)	X	500
4. Absolute error of the mean	α	1.9
5. Average σ/HB ratio range	X ± α	500 ± 2
6. Sample standard deviation	S	8.7
7. Coefficient of variation	V	1.7%
8. Relative error of the mean	ε	0.4%
9. *t* value at 95% two-sided confidence interval for d.f. = 21	$t_{.025}$	2.080
10. Population mean is greater than:	$X-\alpha_x t$	496
11. Population mean is less than:	$X+\alpha_x t$	504

Appendix IV

Hardness and Tensile Strength Relationship for Low-Alloy Steels

Relationship Between Hardness and Tensile Strength

AISI 4027 Low-Alloy Molybdenum Steel
Hardness range (140-320) HB

Brinell hardness, HB (3000-kgf load)	Tensile strength, psi σ	Ratio σ/HB
143	75000	524
156	82000	526
163	86000	528
179	93000	520
179	94000	525
192	95000	495
201	100000	498
201	101000	502
212	104000	491
223	111000	498
229	114000	498
229	114000	498
262	130000	496
285	139000	488
302	144000	477
311	150000	482
321	156000	486
3788	**1888000**	**8532**
average: 223	**average: 111059**	**average: 498**

Relationship Between Hardness and Tensile Strength

AISI 4419 Low-Alloy Molybdenum Steel
Hardness range (140-240) HB

Brinell hardness, HB (3000-kgf load)	Tensile strength, psi σ	Ratio σ/HB
143	72000	503
143	73000	510
143	75000	524
149	78000	523
170	84000	494
179	86000	480
192	92000	479
201	97000	483
201	98500	490
212	102750	485
212	103000	486
217	106500	491
235	118500	504
241	120500	500
2638	**1306750**	**6952**
average: 188	**average: 93339**	**average: 495**

Relationship Between Hardness and Tensile Strength

AISI 4140 Low-Alloy Chromium-molybdenum Steel
Hardness range (190-580) HB

Brinell hardness, HB (3000-kgf load)	Tensile strength, psi σ	Ratio σ/HB
190	95000	500
235	117000	498
255	135000	529
311	148000	476
320	165000	516
341	167000	490
385	200000	519
388	188000	485
429	210000	490
455	230000	505
461	231000	501
495	250000	505
520	260000	500
534	270000	506
578	285000	493
5897	**2951000**	**7513**
average: 393	**average: 196733**	**average: 500**

Relationship Between Hardness and Tensile Strength

AISI 4340 Low-Alloy Nickel-chromium-molybdenum Steel
Hardness range (220-360) HB

Brinell hardness, HB (3000-kgf load)	Tensile strength, psi σ	Ratio σ/HB
217	108000	498
255	124000	486
269	134000	498
269	135000	502
277	139000	502
285	145000	509
293	147000	502
321	161000	502
331	165000	498
331	165000	498
331	166000	502
341	170000	499
341	177000	519
352	175000	497
363	182000	501
363	186000	512
4939	**2479000**	**8025**
average: 309	**average: 154938**	**average: 502**

Statistical Analysis of Data for the Tensile Strength (σ)-Brinell Hardness (HB) Ratio (σ/HB)

AISI 4340 Low-Alloy Steel

Observation	σ/HB	σ/HB-X	$(\sigma/HB-X)^2$
1	498	-2	4
2	486	-14	196
3	498	-2	4
4	502	2	4
5	502	2	4
6	509	9	81
7	502	2	4
8	502	2	4
9	498	-2	4
10	498	-2	4
11	502	2	4
12	499	-1	1
13	519	19	361
14	497	-3	9
15	501	1	1
16	512	12	144
Average ratio: **X** = 500		**Sum of squares:** $\Sigma(\sigma/HB-X)^2 = 829$	

Nomenclature	Symbol	Value
1. Number of observations	**n**	16
2. Number of degrees of freedom (d.f.)	**n - 1**	15
3. Sample mean (average σ/HB ratio)	**X**	500
4. Absolute error of the mean	α	1.9
5. Average σ/HB ratio range	$X \pm \alpha$	500 ± 2
6. Sample standard deviation	**S**	7.4
7. Coefficient of variation	**V**	1.5%
8. Relative error of the mean	ε	0.4%
9. *t* value at 95% two-sided confidence interval for d.f. = 15	$t_{.025}$	2.131
10. Population mean is greater than:	$X-\alpha_x t$	496
11. Population mean is less than:	$X+\alpha_x t$	504

Relationship Between Hardness and Tensile Strength

AISI 4340 Low-Alloy Steel, Hardness range (390-600) HB

Brinell hardness, HB (3000-kgf load)	Tensile strength, psi σ	Ratio σ/HB
388	210000	541
498	269000	540
514	281000	547
600	319000	532
2000	**1079000**	**2160**
average: 500	**average: 269750**	**average: 540**

Relationship Between Hardness and Tensile Strength

AISI 8620 Low-Alloy Nickel-chromium-molybdenum Steel
Hardness range (150-390) HB

Brinell hardness, HB (3000-kgf load)	Tensile strength, psi σ	Ratio σ/HB
149	77000	517
149	78000	523
163	82000	503
179	87000	486
183	92000	503
197	96000	487
201	98000	488
207	98000	473
229	114000	498
235	117000	498
248	124000	500
255	127000	498
352	178000	506
388	200000	515
3135	**1568000**	**6995**
average: 224	**average: 112000**	**average: 500**

Statistical Analysis of Data for the Tensile Strength (σ)-Brinell Hardness (HB) Ratio (σ/HB)

AISI 8620 Low-Alloy Steel

Observation	σ/HB	σ/HB-X	$(\sigma/HB-X)^2$
1	517	17	289
2	523	23	529
3	503	3	9
4	486	-14	196
5	503	3	9
6	487	-13	169
7	488	-12	144
8	473	-27	729
9	498	-2	4
10	498	-2	4
11	500	0	0
12	498	-2	4
13	506	6	36
14	515	15	225
Average ratio: X = 500		**Sum of squares:** $\Sigma(\sigma/HB-X)^2 = 2347$	

Nomenclature	Symbol	Value
1. Number of observations	n	14
2. Number of degrees of freedom (d.f.)	n - 1	13
3. Sample mean (average σ/HB ratio)	X	500
4. Absolute error of the mean	α	3.6
5. Average σ/HB ratio range	$X \pm \alpha$	500 ± 4
6. Sample standard deviation	S	13.4
7. Coefficient of variation	V	2.7%
8. Relative error of the mean	ε	0.7%
9. *t* value at 95% two-sided confidence interval for d.f. = 13	$t_{.025}$	2.160
10. Population mean is greater than:	$X-\alpha_x t$	492
11. Population mean is less than:	$X+\alpha_x t$	508

Relationship Between Hardness and Tensile Strength

AISI 8630 Low-Alloy Nickel-chromium-molybdenum Steel
Hardness range (160-300) HB

Brinell hardness, HB (3000-kgf load)	Tensile strength, psi σ	Ratio σ/HB
156	82000	526
156	82000	526
187	92000	492
187	93000	497
187	94000	503
187	94000	503
197	96000	487
217	107000	493
223	111000	498
235	113000	481
235	120000	511
241	118000	490
269	130000	483
269	134000	498
269	135000	502
285	139000	488
293	147000	502
302	152000	503
4095	**2039000**	**8983**
average: 228	**average: 113278**	**average: 498**

Relationship Between Hardness and Tensile Strength

AISI 8650 Low-Alloy Nickel-chromium-molybdenum Steel
Hardness range (210-360) HB

Brinell hardness, HB (3000-kgf load)	Tensile strength, psi σ	Ratio σ/HB
212	104000	491
241	122000	506
255	126000	494
277	135000	487
285	139000	488
285	141000	495
285	143000	502
293	144000	491
293	145000	495
293	148000	505
302	148000	490
302	149000	493
311	154000	495
331	165000	498
352	172000	489
363	178000	490
363	182000	501
5043	**2495000**	**8410**
average: 297	**average: 146765**	**average: 495**

Relationship Between Hardness and Tensile Strength

AISI 8740 Low-Alloy Nickel-chromium-molybdenum Steel
Hardness range (200-350) HB

Brinell hardness, HB (3000-kgf load)	Tensile strength, psi σ	Ratio σ/HB
201	101000	502
229	116000	507
248	124000	500
255	127000	498
255	132000	518
262	132000	504
269	135000	502
269	136000	506
277	139000	502
277	142000	513
285	138000	484
285	140000	491
302	149000	493
311	154000	495
331	171000	517
352	178000	506
352	179000	509
4760	**2393000**	**8547**
average: 280	**average: 140765**	**average: 503**

Relationship Between Hardness and Tensile Strength

AISI 9310 Low-Alloy Nickel-chromium-molybdenum Steel
Hardness range (240-375) HB

Brinell hardness, HB (3000-kgf load)	Tensile strength, psi σ	Ratio σ/HB
241	119000	494
255	125000	490
262	131000	500
269	132000	491
269	132000	491
277	136000	491
293	144000	491
293	145000	495
321	158000	492
321	159000	495
341	168000	493
363	179000	493
375	187000	499
3880	**1915000**	**6415**
average: 298	**average: 147308**	**average: 494**

Relationship Between Hardness and Tensile Strength

AISI 4620 Low-Alloy Nickel-molybdenum Steel
Hardness range (150-255) HB

Brinell hardness, HB (3000-kgf load)	Tensile strength, psi σ	Ratio σ/HB
149	74000	497
167	80000	479
170	84000	494
170	85000	500
174	83000	477
187	96000	513
192	96000	500
192	98000	510
197	98000	497
241	118000	490
255	127000	498
2094	**1039000**	**5455**
average: 190	**average: 94455**	**average: 496**

Relationship Between Hardness and Tensile Strength

AISI 4820 Low-Alloy Nickel-molybdenum Steel
Hardness range (200-400) HB

Brinell hardness, HB (3000-kgf load)	Tensile strength, psi σ	Ratio σ/HB
197	99000	503
212	104000	491
223	107000	480
229	110000	480
235	112000	477
235	117000	498
241	119000	494
269	130000	483
277	136000	491
331	163000	492
352	170000	483
388	205000	528
401	209000	521
3590	**1781000**	**6421**
average: 276	**average: 137000**	**average: 496**

Relationship Between Hardness and Tensile Strength

AISI 5140 Low-Alloy Chromium Steel
Hardness range (170-300) HB

Brinell hardness, HB (3000-kgf load)	Tensile strength, psi σ	Ratio σ/HB
167	83000	497
167	83000	497
217	106000	488
217	111000	512
223	110000	493
223	113000	507
229	115000	502
229	115000	502
235	116000	494
235	117000	498
235	120000	511
241	118000	490
241	120000	498
248	125000	504
255	128000	502
262	127000	485
269	130000	483
293	141000	481
302	147000	487
4488	**2225000**	**9431**
average: 236	**average: 117105**	**average: 496**

Relationship Between Hardness and Tensile Strength

AISI 5150 Low-Alloy Chromium Steel
Hardness range (200-310) HB

Brinell hardness, HB (3000-kgf load)	Tensile strength, psi σ	Ratio σ/HB
197	98000	497
197	98000	497
235	115000	489
241	119000	494
241	120000	498
241	122000	506
248	123000	496
248	125000	504
255	126000	494
255	126000	494
255	127000	498
255	128000	502
255	132000	518
262	131000	500
269	136000	506
277	137000	495
285	144000	505
302	153000	507
311	159000	511
4829	**2419000**	**9511**
average: 254	**average: 127316**	**average: 501**

Relationship Between Hardness and Tensile Strength

AISI 5160 Low-Alloy Chromium Steel
Hardness range (240-340) HB

Brinell hardness, HB (3000-kgf load)	Tensile strength, psi σ	Ratio σ/HB
241	120000	498
255	134000	525
262	129000	492
262	129000	492
262	134000	511
269	133000	494
269	139000	517
277	135000	487
285	140000	491
285	149000	523
293	154000	526
302	145000	480
302	152000	503
341	166000	487
341	170000	499
4246	**2129000**	**7525**
average: 283	**average: 141933**	**average: 501**

Relationship Between Hardness and Tensile Strength

AISI 6150 Low-Alloy Chromium-vanadium Steel
Hardness range (200-360) HB

Brinell hardness, HB (3000-kgf load)	Tensile strength, psi σ	Ratio σ/HB
197	97000	492
241	122000	506
255	128000	502
262	130000	496
262	130000	496
269	134000	498
269	136000	506
285	141000	495
293	141000	481
293	147000	502
293	148000	505
302	152000	503
311	158000	508
321	160000	498
331	166000	502
352	174000	494
363	180000	496
4899	**2444000**	**8480**
average: 288	**average: 143765**	**average: 499**

Statistical Analysis of Data for the Tensile Strength (σ)-Brinell Hardness (HB) Ratio (σ/HB)

AISI 6150 Low-Alloy Steel

Observation	σ/HB	σ/HB-X	$(\sigma/HB-X)^2$
1	492	-8	64
2	506	6	36
3	502	2	4
4	496	-4	16
5	496	-4	16
6	498	-2	4
7	506	6	36
8	495	-5	25
9	481	-19	361
10	502	2	4
11	505	5	25
12	503	3	9
13	508	8	64
14	498	-2	4
15	502	2	4
16	494	-6	36
17	496	-4	16
Average ratio: **X** = 500		**Sum of squares:** $\Sigma(\sigma/HB-X)^2 = 724$	

Nomenclature	Symbol	Value
1. Number of observations	**n**	17
2. Number of degrees of freedom (d.f.)	**n - 1**	16
3. Sample mean (average σ/HB ratio)	**X**	500
4. Absolute error of the mean	α	1.6
5. Average σ/HB ratio range	$X \pm \alpha$	500 ± 2
6. Sample standard deviation	**S**	6.7
7. Coefficient of variation	**V**	1.3%
8. Relative error of the mean	ε	0.3%
9. *t* value at 95% two-sided confidence interval for d.f. = 16	$t_{.025}$	2.120
10. Population mean is greater than:	$X-\alpha_x\, t$	497
11. Population mean is less than:	$X+\alpha_x\, t$	503

Relationship Between Hardness and Tensile Strength

AISI 9255 Low-Alloy Silicon-manganese Steel
Hardness range (230-330) HB

Brinell hardness, HB (3000-kgf load)	Tensile strength, psi σ	Ratio σ/HB
229	113000	493
262	132000	504
269	133000	494
269	135000	502
269	135000	502
277	137000	495
277	138000	498
277	138000	498
277	138000	498
285	145000	509
293	146000	498
293	149000	509
293	150000	512
302	155000	513
302	155000	513
321	164000	511
331	170000	514
4826	**2433000**	**8563**
average: 284	**average: 143118**	**average: 504**

Appendix V

Hardness and Tensile Strength Relationship for Stainless Steels

TENSILE STRENGTH versus BRINELL HARDNESS

AISI Type 201 austenitic stainless steel, Hardness range (250-400) HB

$$\sigma = 606 \times HB - 31660$$

Brinell hardness	Tensile strength, psi	*XY*	X^2	Y^2	Parameters of the regression line		Correlation coefficient
HB, X	***σ, Y***				***slope***	***intercept***	
253	125000	31625000	64009	1.563E+10	606	-31660	0.962
301	150000	45150000	90601	2.25E+10			
344	175000	60200000	118336	3.063E+10			
381	185000	70485000	145161	3.423E+10			
400	225000	90000000	160000	5.063E+10			
1679	860000	297460000	578107	1.536E+11			

Relationship Between Hardness and Tensile Strength

300 series austenitic stainless steels, Hardness range (140-180) HB

AISI type	Brinell hardness, HB (3000-kgf load)	Tensile strength, psi σ	Ratio σ/HB
302B (plate)	165	90000	545
302B (bar)	165	90000	545
303, 303Se	160	90000	563
303MA	160	90000	563
304 (plate)	149	82000	550
304 (bar)	149	85000	570
308 (plate)	150	85000	567
308 (bar)	150	85000	567
309, 309S (bar)	159	95000	597
309, 309S (plate)	170	95000	559
310, 310S (bar)	160	95000	594
310, 310S (plate)	170	95000	559
314 (plate)	180	100000	556
314 (bar)	180	100000	556
316 (bar)	149	80000	537
316 (plate)	149	82000	550
316L (plate)	146	81000	555
316F (bar)	143	82000	573
317 (plate)	160	85000	531
317 (bar)	160	85000	531
317L (plate)	150	85000	567
321 (bar)	150	85000	567
321 (plate)	160	85000	531
330 (bar)	150	85000	567
330 (plate)	150	90000	600
347, 348 (plate)	160	90000	563
347, 348 (bar)	160	90000	563
	4254	**2382000**	**15126**
	average: 158	**average: 88222**	**average: 560**

Statistical Analysis of Data for the Tensile Strength (σ)-Brinell Hardness (HB) Ratio (σ/HB)

300 series austenitic stainless steels, Hardness range (140-180) HB

Observation	σ/HB	σ/HB-X	$(\sigma/HB-X)^2$
1	545	-15	225
2	545	-15	225
3	563	3	9
4	563	3	9
5	550	-10	100
6	570	10	100
7	567	7	49
8	567	7	49
9	597	37	1369
10	559	-1	1
11	594	34	1156
12	559	-1	1
13	556	-4	16
14	556	-4	16
15	537	-23	529
16	550	-10	100
17	555	-5	25
18	573	13	169
19	531	-29	841
20	531	-29	841
21	567	7	49
22	567	7	49
23	531	-29	841
24	567	7	49
25	600	40	1600
26	563	3	9
27	563	3	9
Average ratio: X = 560		**Sum of squares:** $\Sigma(\sigma/HB-X)^2 = 8436$	

Nomenclature	Symbol	Value
1. Number of observations	**n**	27
2. Number of degrees of freedom (d.f.)	**n - 1**	26
3. Sample mean (average σ/HB ratio)	**X**	560
4. Absolute error of the mean	α	3.46
5. Average σ/HB ratio range	$X \pm \alpha$	560 ± 3
6. Sample standard deviation	**S**	18.0
7. Coefficient of variation	**V**	3.2%
8. Relative error of the mean	ε	0.6%
9. *t* value at 95% two-sided confidence interval for d.f. = 26	$t_{.025}$	2.056
10. Population mean is greater than:	$X-\alpha_x t$	553
11. Population mean is less than:	$X+\alpha_x t$	567

TENSILE STRENGTH versus BRINELL HARDNESS

300 series austenitic stainless steels, Hardness range (140-180) HB

$$\sigma = 457 \times HB + 16910$$

Brinell hardness	Tensile strength, psi	*XY*	X^2	Y^2	Parameters of the regression line		Correlation coefficient
HB, X	*σ, Y*				*slope*	*intercept*	
143	82000	11726000	20449	6.724E+09	457	16910	0.962
149	85000	12665000	22201	7.225E+09			
150	85000	12750000	22500	7.225E+09			
150	85000	12750000	22500	7.225E+09			
150	85000	12750000	22500	7.225E+09			
150	85000	12750000	22500	7.225E+09			
150	85000	12750000	22500	7.225E+09			
150	90000	13500000	22500	8.1E+09			
160	90000	14400000	25600	8.1E+09			
160	90000	14400000	25600	8.1E+09			
160	90000	14400000	25600	8.1E+09			
160	90000	14400000	25600	8.1E+09			
165	90000	14850000	27225	8.1E+09			
165	90000	14850000	27225	8.1E+09			
170	95000	16150000	28900	9.025E+09			
170	95000	16150000	28900	9.025E+09			
180	100000	18000000	32400	1E+10			
180	100000	18000000	32400	1E+10			
2862	1612000	257241000	457100	1.448E+11			

Relationship Between Hardness and Tensile Strength

300 series austenitic stainless steels, Hardness range (190-370) HB

AISI type	Brinell hardness, HB (3000-kgf load)	Tensile strength, psi σ	Ratio σ/HB
301, 25% hard	253	125000	494
301, Full hard	371	185000	499
303, 303Se (bar)	228	100000	439
303MA (bar)	228	100000	439
304 (bar)	212	100000	472
305 (bar)	240	110000	458
306 (bar)	277	125000	451
316 (bar)	190	90000	474
316L (bar)	190	85000	447
329 (bar)	230	105000	457
347, 348 (bar)	212	100000	472
	2631	**1225000**	**5102**
	average: 239	**average: 111364**	**average: 466**

Statistical Analysis of Data for the Tensile Strength (σ)-Brinell Hardness (HB) Ratio (σ/HB)

300 series austenitic stainless steels, Hardness range (190-370) HB

Observation	σ/HB	σ/HB-X	$(\sigma/HB-X)^2$
1	494	24	576
2	499	29	841
3	439	-31	961
4	439	-31	961
5	472	2	4
6	458	-12	144
7	451	-19	361
8	474	4	16
9	447	-23	529
10	457	-13	169
11	472	2	4
Average ratio: **X** = 470		**Sum of squares:** $\Sigma(\sigma/HB-X)^2 = 4566$	

Nomenclature	Symbol	Value
1. Number of observations	**n**	11
2. Number of degrees of freedom (d.f.)	**n - 1**	10
3. Sample mean (average σ/HB ratio)	**X**	470
4. Absolute error of the mean	α	6.45
5. Average σ/HB ratio range	$X \pm \alpha$	470 ± 6
6. Sample standard deviation	**S**	21.4
7. Coefficient of variation	**V**	4.6%
8. Relative error of the mean	ε	1.4%
9. *t* value at 95% two-sided confidence interval for d.f. = 10	$t_{.025}$	2.228
10. Population mean is greater than:	$X-\alpha_x t$	456
11. Population mean is less than:	$X+\alpha_x t$	484

TENSILE STRENGTH versus BRINELL HARDNESS

300 series austenitic stainless steels, Hardness range (190-370) HB

$$\sigma = 534 \times HB - 16280$$

Brinell hardness	Tensile strength, psi	*XY*	X^2	Y^2	Parameters of the regression line		Correlation coefficient
HB, X	***σ, Y***				***slope***	***intercept***	
190	85000	16150000	36100	7.225E+09	534	-16280	0.987
190	90000	17100000	36100	8.1E+09			
212	100000	21200000	44944	1E+10			
212	100000	21200000	44944	1E+10			
228	100000	22800000	51984	1E+10			
228	100000	22800000	51984	1E+10			
230	105000	24150000	52900	1.103E+10			
240	110000	26400000	57600	1.21E+10			
253	125000	31625000	64009	1.563E+10			
277	125000	34625000	76729	1.563E+10			
371	185000	68635000	137641	3.423E+10			
2631	1225000	306685000	654935	1.439E+11			

Relationship Between Hardness and Tensile Strength

400 series ferritic stainless steels, Hardness range (140-190) HB

AISI type	Brinell hardness, HB (3000-kgf load)	Tensile strength, psi σ	Ratio σ/HB
405 (bar)	150	70000	467
405 (bar)	185	85000	459
409 (plate, bar)	137	65000	474
430F (bar)	170	80000	471
430F (bar)	190	90000	474
430, 434, 435, 436	160	75000	469
430, 434, 435, 436	155	75000	484
430, 434, 435, 436	185	85000	459
442, 443 (bar)	160	75000	469
446 (bar)	169	80000	473
446 (bar)	185	85000	459
	1846	**865000**	**5158**
	average: 168	**average: 78636**	**average: 469**

TENSILE STRENGTH versus BRINELL HARDNESS

400 series ferritic stainless steels, Hardness range (140-190) HB

$$\sigma = 430 \times HB + 6530$$

Brinell hardness	Tensile strength, psi	*XY*	X^2	Y^2	Parameters of the regression line		Correlation coefficient
HB, X	*σ, Y*				*slope*	*intercept*	
137	65000	8905000	18769	4.225E+09	430	6530	0.989
150	70000	10500000	22500	4.9E+09			
155	75000	11625000	24025	5.625E+09			
160	75000	12000000	25600	5.625E+09			
160	75000	12000000	25600	5.625E+09			
169	80000	13520000	28561	6.4E+09			
170	80000	13600000	28900	6.4E+09			
185	85000	15725000	34225	7.225E+09			
185	85000	15725000	34225	7.225E+09			
185	85000	15725000	34225	7.225E+09			
190	90000	17100000	36100	8.1E+09			
1846	865000	146425000	312730	6.858E+10			

Relationship Between Hardness and Tensile Strength

400 series martensitic stainless steels, Hardness range (156-595) HB

AISI type	Brinell hardness, HB (3000-kgf load)	Tensile strength, psi σ	Ratio σ/HB
403 (bar)	156	75000	481
403 (bar)	222	111000	500
410, 410S (bar)	421	221000	525
414 (plate)	235	115000	489
414 (bar)	235	115000	489
414 (bar)	270	130000	481
416, 416Se (bar)	156	82000	526
416, 416Se (bar)	205	100000	488
416, 416Se (bar)	222	110000	495
416, 416Se (bar)	226	107500	476
416, 416Se (bar)	245	115500	471
416, 416Se (bar)	258	125500	486
416, 416Se (bar)	319	157000	492
416, 416Se (bar)	362	172000	475
416, 416Se (bar)	362	182500	504
416, 416Se (bar)	362	187500	518
416, 416Se (bar)	366	190000	519
420 (bar)	195	95000	487
420 (bar)	215	105000	488
420 (bar)	500	250000	500
420 (bar)	595	294000	494
420F (bar)	228	110000	482
431 (bar)	260	125000	481
431 (bar)	270	130000	481
431 (bar)	421	208000	494
440A (bar)	210	105000	500
440A (bar)	235	115000	489
440A (bar)	510	260000	510
440B (bar)	216	107000	495
440B (bar)	243	120000	494
440B (bar)	555	280000	505
440C (bar)	222	110000	495
440C (bar)	260	125000	481
440C (bar)	580	285000	491
	10337	**5120500**	**16782**
	average: 304	**average: 150603**	**average: 495**

TENSILE STRENGTH versus BRINELL HARDNESS

400 series martensitic stainless steels, Hardness range (156-595) HB

$$\sigma = 508 \times HB - 3900$$

Brinell hardness	Tensile strength, psi	XY	X^2	Y^2	Parameters of the regression line		Correlation coefficient
HB, X	σ, Y				slope	intercept	
156	75000	11700000	24336	5625000000	508	-3900	0.998
156	82000	12792000	24336	6724000000			
195	95000	18525000	38025	9025000000			
205	100000	20500000	42025	1E+10			
210	105000	22050000	44100	1.1025E+10			
215	105000	22575000	46225	1.1025E+10			
216	107000	23112000	46656	1.1449E+10			
222	110000	24420000	49284	1.21E+10			
222	110000	24420000	49284	1.21E+10			
222	111000	24642000	49284	1.2321E+10			
226	107500	24295000	51076	1.1556E+10			
228	110000	25080000	51984	1.21E+10			
235	115000	27025000	55225	1.3225E+10			
235	115000	27025000	55225	1.3225E+10			
235	115000	27025000	55225	1.3225E+10			
243	120000	29160000	59049	1.44E+10			
245	115500	28297500	60025	1.334E+10			
258	125500	32379000	66564	1.575E+10			
260	125000	32500000	67600	1.5625E+10			
260	125000	32500000	67600	1.5625E+10			
270	130000	35100000	72900	1.69E+10			
270	130000	35100000	72900	1.69E+10			
319	157000	50083000	101761	2.4649E+10			
362	172000	62264000	131044	2.9584E+10			
362	182500	66065000	131044	3.3306E+10			
362	187500	67875000	131044	3.5156E+10			
366	190000	69540000	133956	3.61E+10			
421	208000	87568000	177241	4.3264E+10			
421	221000	93041000	177241	4.8841E+10			
500	250000	125000000	250000	6.25E+10			
510	260000	132600000	260100	6.76E+10			
555	280000	155400000	308025	7.84E+10			
580	285000	165300000	336400	8.1225E+10			
595	294000	174930000	354025	8.6436E+10			
10337	5120500	1809888500	3640809	9.0033E+11			

Relationship Between Hardness and Tensile Strength

Precipitation-hardening stainless steels, Hardness range (250-430) HB

Designation	Brinell hardness, HB (3000-kgf load)	Tensile strength, psi σ	Ratio σ/HB
PH 13-8 Mo	258	125000	484
PH 13-8 Mo	286	135000	472
PH 13-8 Mo	319	150000	470
PH 13-8 Mo	371	175000	472
PH 13-8 Mo	381	185000	486
PH 13-8 Mo	400	200000	500
PH 13-8 Mo	400	205000	513
PH 13-8 Mo	421	220000	523
PH 13-8 Mo	421	220000	523
15-5 PH	271	135000	498
15-5 PH	279	140000	502
15-5 PH	294	140000	476
15-5 PH	301	145000	482
15-5 PH	327	155000	474
15-5 PH	353	170000	482
15-5 PH	371	190000	512
15-5 PH	371	190000	512
PH 15-7 Mo	371	190000	512
PH 15-7 Mo	381	200000	525
PH 15-7 Mo	432	225000	521
17-4 PH	271	135000	498
17-4 PH	279	140000	502
17-4 PH	294	140000	476
17-4 PH	301	145000	482
17-4 PH	327	155000	474
17-4 PH	353	170000	482
17-4 PH	371	190000	512
17-4 PH	371	190000	512
17-7 PH	353	170000	482
17-7 PH	353	180000	510
17-7 PH	381	185000	486
17-7 PH	381	200000	525
17-7 PH	400	210000	525
AM-350	336	165000	491
AM-350	390	185000	474
AM-355	344	170000	494
AM-355	344	170000	494
Custom 450	253	130000	514
Custom 450	258	125000	484
Custom 450	336	160000	476
Custom 450	371	180000	485
	14075	**6990000**	**20317**
	average: 343	**average: 170488**	**average: 497**

TENSILE STRENGTH versus BRINELL HARDNESS

Precipitation-hardening stainless steels, Hardness range (250-430) HB

$$\sigma = 557 \times HB - 20710$$

Brinell hardness HB, X	Tensile strength, psi σ, Y	XY	X^2	Y^2	Parameters of the regression line slope	intercept	Correlation coefficient
253	130000	32890000	64009	1.69E+10	557	-20710	0.979
258	125000	32250000	66564	1.563E+10			
258	125000	32250000	66564	1.563E+10			
271	135000	36585000	73441	1.823E+10			
271	135000	36585000	73441	1.823E+10			
279	140000	39060000	77841	1.96E+10			
279	140000	39060000	77841	1.96E+10			
286	135000	38610000	81796	1.823E+10			
294	140000	41160000	86436	1.96E+10			
294	140000	41160000	86436	1.96E+10			
301	145000	43645000	90601	2.103E+10			
301	145000	43645000	90601	2.103E+10			
319	150000	47850000	101761	2.25E+10			
327	155000	50685000	106929	2.403E+10			
327	155000	50685000	106929	2.403E+10			
336	160000	53760000	112896	2.56E+10			
336	165000	55440000	112896	2.723E+10			
344	170000	58480000	118336	2.89E+10			
344	170000	58480000	118336	2.89E+10			
353	170000	60010000	124609	2.89E+10			
353	170000	60010000	124609	2.89E+10			
353	170000	60010000	124609	2.89E+10			
353	180000	63540000	124609	3.24E+10			
371	175000	64925000	137641	3.063E+10			
371	180000	66780000	137641	3.24E+10			
371	190000	70490000	137641	3.61E+10			
371	190000	70490000	137641	3.61E+10			
371	190000	70490000	137641	3.61E+10			
371	190000	70490000	137641	3.61E+10			
371	190000	70490000	137641	3.61E+10			
381	185000	70485000	145161	3.423E+10			
381	185000	70485000	145161	3.423E+10			
381	200000	76200000	145161	4E+10			
381	200000	76200000	145161	4E+10			
390	185000	72150000	152100	3.423E+10			
400	200000	80000000	160000	4E+10			
400	205000	82000000	160000	4.203E+10			
400	210000	84000000	160000	4.41E+10			
421	220000	92620000	177241	4.84E+10			
421	220000	92620000	177241	4.84E+10			
432	225000	97200000	186624	5.063E+10			
14075	6990000	2.454E+09	4929427	1.223E+12			

TENSILE STRENGTH versus ROCKWELL HARDNESS "C" SCALE

Precipitation-hardening stainless steels, Hardness range (25-46) HRC

$$\sigma = 4725 \times HRC - 2370$$

Rockwell hardness	Tensile strength, psi	*XY*	*X²*	*Y²*	Parameters of the regression line		Correlation coefficient
HRC, X	***σ, Y***				***slope***	***intercept***	
25	130000	3250000	625	1.69E+10	4725	-2370	0.970
26	125000	3250000	676	1.563E+10			
26	125000	3250000	676	1.563E+10			
28	135000	3780000	784	1.823E+10			
28	135000	3780000	784	1.823E+10			
29	140000	4060000	841	1.96E+10			
29	140000	4060000	841	1.96E+10			
30	135000	4050000	900	1.823E+10			
31	140000	4340000	961	1.96E+10			
31	140000	4340000	961	1.96E+10			
32	145000	4640000	1024	2.103E+10			
32	145000	4640000	1024	2.103E+10			
34	150000	5100000	1156	2.25E+10			
35	155000	5425000	1225	2.403E+10			
35	155000	5425000	1225	2.403E+10			
36	160000	5760000	1296	2.56E+10			
36	165000	5940000	1296	2.723E+10			
37	170000	6290000	1369	2.89E+10			
37	170000	6290000	1369	2.89E+10			
38	170000	6460000	1444	2.89E+10			
38	170000	6460000	1444	2.89E+10			
38	170000	6460000	1444	2.89E+10			
38	180000	6840000	1444	3.24E+10			
40	175000	7000000	1600	3.063E+10			
40	180000	7200000	1600	3.24E+10			
40	190000	7600000	1600	3.61E+10			
40	190000	7600000	1600	3.61E+10			
40	190000	7600000	1600	3.61E+10			
40	190000	7600000	1600	3.61E+10			
40	190000	7600000	1600	3.61E+10			
41	185000	7585000	1681	3.423E+10			
41	185000	7585000	1681	3.423E+10			
41	200000	8200000	1681	4E+10			
41	200000	8200000	1681	4E+10			
42	185000	7770000	1764	3.423E+10			
43	200000	8600000	1849	4E+10			
43	205000	8815000	1849	4.203E+10			
43	210000	9030000	1849	4.41E+10			
45	220000	9900000	2025	4.84E+10			
45	220000	9900000	2025	4.84E+10			
46	225000	10350000	2116	5.063E+10			
1500	6990000	262025000	56210	1.223E+12			

Relationship Between Hardness and Tensile Strength

Precipitation-hardening stainless steels, Hardness range (370-440) HB

Designation and condition*	Brinell hardness, HB (3000-kgf load)	Tensile strength, psi σ	Ratio σ/HB
Custom 455, H1000	371	205000	553
Custom 455, H950	409	220000	538
Custom 455, H950	409	222000	543
Custom 455, H900	443	235000	530
PH 15-7 Mo, CH900	432	240000	556
17-7 PH, CH900	432	240000	556
	2496	**1362000**	**3276**
	average: 416	**average: 227000**	**average: 546**

* Condition:
H, hardened; CH, cold rolled and hardened.
Numerical code shows the aging temperature in °F

Appendix VI

Hardness and Tensile Strength Relationship for Tool Steels

TENSILE STRENGTH versus ROCKWELL HARDNESS "C" SCALE

AISI Type S1 shock-resisting tool steel, Oil quenched, Hardness range (42-58) HRC

$$\sigma = 7014 \times HRC - 94200$$

Rockwell hardness	Tensile strength, psi	XY	X^2	Y^2	Parameters of the regression line		Correlation coefficient
HRC, X	**σ, Y**				***slope***	***intercept***	
42	195000	8190000	1764	3.803E+10	7014	-94200	0.984
47.5	244000	11590000	2256.25	5.954E+10			
50.5	260000	13130000	2550.25	6.76E+10			
54	294000	15876000	2916	8.644E+10			
57.5	300000	17250000	3306.25	9E+10			
251.5	1293000	66036000	12792.75	3.416E+11			

TENSILE STRENGTH versus ROCKWELL HARDNESS "C" SCALE

AISI Type S5 shock-resisting tool steel, Oil quenched, Hardness range (37-59) HRC

$$\sigma = 7829 \times HRC - 131700$$

Rockwell hardness	Tensile strength, psi	*XY*	*X²*	*Y²*	Parameters of the regression line		Correlation coefficient
HRC, X	***σ, Y***				***slope***	***intercept***	
37	170000	6290000	1369	2.89E+10	7829	-131700	0.980
48	220000	10560000	2304	4.84E+10			
52	275000	14300000	2704	7.563E+10			
58	325000	18850000	3364	1.056E+11			
59	340000	20060000	3481	1.156E+11			
254	1330000	70060000	13222	3.742E+11			

TENSILE STRENGTH versus ROCKWELL HARDNESS "C" SCALE

AISI Type S7 shock-resisting tool steel, Oil quenched, Hardness range (39-58) HRC

$$\sigma = 6923 \times HRC - 92700$$

Rockwell hardness	Tensile strength, psi	*XY*	X^2	Y^2	Parameters of the regression line		Correlation coefficient
HRC, X	***σ, Y***				***slope***	***intercept***	
39	180000	7020000	1521	3.24E+10	6923	-92700	0.995
51	254000	12954000	2601	6.452E+10			
53	275000	14575000	2809	7.563E+10			
55	285000	15675000	3025	8.123E+10			
58	315000	18270000	3364	9.923E+10			
256	1309000	68494000	13320	3.53E+11			

TENSILE STRENGTH versus ROCKWELL HARDNESS "C" SCALE

AISI Type 01 oil-hardening tool steel, Oil quenched, Hardness range (31-50) HRC

$$\sigma = 5899 \times HRC - 60120$$

Rockwell hardness HRC, X	Tensile strength, psi σ, Y	XY	X^2	Y^2	Parameters of the regression line: slope	Parameters of the regression line: intercept	Correlation coefficient
31	133000	4123000	961	1.769E+10	5899	-60120	0.966
39	159000	6201000	1521	2.528E+10			
44	187000	8228000	1936	3.497E+10			
47	217000	10199000	2209	4.709E+10			
50	248000	12400000	2500	6.15E+10			
211	944000	41151000	9127	1.865E+11			

TENSILE STRENGTH versus ROCKWELL HARDNESS "C" SCALE

AISI Type O7 oil-hardening tool steel, Oil quenched, Hardness range (31-54) HRC

$$\sigma = 6332 \times HRC - 78970$$

Rockwell hardness	Tensile strength, psi	XY	X^2	Y^2	Parameters of the regression line		Correlation coefficient
HRC, X	***σ, Y***				***slope***	***intercept***	
30.9	128000	3955200	954.81	1.638E+10	6332	-78970	0.983
40.2	161000	6472200	1616.04	2.592E+10			
46.8	209000	9781200	2190.24	4.368E+10			
50.4	237000	11944800	2540.16	5.617E+10			
52.1	258000	13441800	2714.41	6.656E+10			
54.2	272000	14742400	2937.64	7.398E+10			
274.6	1265000	60337600	12953.30	2.827E+11			

TENSILE STRENGTH versus ROCKWELL HARDNESS "C" SCALE

AISI Type A2 air-hardening tool steel, Air cooled, Hardness range (30-60) HRC

$$\sigma = 5920 \times HRC - 53270$$

Rockwell hardness	Tensile strength, psi	*XY*	X^2	Y^2	Parameters of the regression line		Correlation coefficient
HRC, X	***σ, Y***				***slope***	***intercept***	
30	128500	3855000	900	1.65E+10	5920	-53270	0.999
35	152300	5330500	1225	2.32E+10			
40	181300	7252000	1600	3.29E+10			
45	211300	9508500	2025	4.46E+10			
50	241400	12070000	2500	5.83E+10			
55	272500	14987500	3025	7.43E+10			
60	304600	18276000	3600	9.28E+10			
315	1491900	71279500	14875	3.43E+11			

TENSILE STRENGTH versus ROCKWELL HARDNESS "C" SCALE

AISI Type H11, H13 chromium hot-work tool steels, Air or oil quenched, Hardness range (30-57) HRC

$$\sigma = 6387 \times HRC - 62850$$

Rockwell hardness	Tensile strength, psi	*XY*	X^2	Y^2	Parameters of the regression line		Correlation coefficient
HRC, X	***σ, Y***				***slope***	***intercept***	
30	138100	4143000	900	1.9072E+10	6387	-62850	0.994
35	160600	5621000	1225	2.5792E+10			
40	185300	7412000	1600	3.4336E+10			
45	216000	9720000	2025	4.6656E+10			
50	252100	12605000	2500	6.3554E+10			
55	292500	16087500	3025	8.5556E+10			
57	308200	17567400	3249	9.4987E+10			
312	1552800	73155900	14524	3.6995E+11			

TENSILE STRENGTH versus ROCKWELL HARDNESS "C" SCALE

AISI Type L2 special-purpose tool steel, Oil quenched, Hardness range (30-54) HRC

$$\sigma = 6227 \times HRC - 59980$$

Rockwell hardness	Tensile strength, psi	*XY*	X^2	Y^2	Parameters of the regression line		Correlation coefficient
HRC, X	***σ, Y***				***slope***	***intercept***	
30	135000	4050000	900	1.823E+10	6227	-59980	0.985
41	185000	7585000	1681	3.423E+10			
47	225000	10575000	2209	5.063E+10			
52	260000	13520000	2704	6.76E+10			
54	290000	15660000	2916	8.41E+10			
224	1095000	51390000	10410.00	2.548E+11			

TENSILE STRENGTH versus ROCKWELL HARDNESS "C" SCALE

AISI Type L6 special-purpose tool steel, Oil quenched, Hardness range (32-54) HRC

$$\sigma = 6843 \times HRC - 83900$$

Rockwell hardness	Tensile strength, psi	*XY*	X^2	Y^2	Parameters of the regression line		Correlation coefficient
HRC, X	***σ, Y***				***slope***	***intercept***	
32	140000	4480000	1024	1.96E+10	6843	-83900	0.995
42	195000	8190000	1764	3.803E+10			
46	230000	10580000	2116	5.29E+10			
54	290000	15660000	2916	8.41E+10			
174	855000	38910000	7820	1.946E+11			

TENSILE STRENGTH versus ROCKWELL HARDNESS "C" SCALE

AISI Type P20 mold tool steel, Oil quenched, Hardness range (26-54) HRC

$$\sigma = 5535 \times HRC - 24880$$

Rockwell hardness	Tensile strength, psi	*XY*	X^2	Y^2	Parameters of the regression line		Correlation coefficient
HRC, X	***σ, Y***				***slope***	***intercept***	
26	130000	3380000	676	1.69E+10	5535	-24880	0.990
30	144000	4320000	900	2.074E+10			
35	168000	5880000	1225	2.822E+10			
38	180000	6840000	1444	3.24E+10			
41	192000	7872000	1681	3.686E+10			
44	207000	9108000	1936	4.285E+10			
46	224000	10304000	2116	5.018E+10			
48	242000	11616000	2304	5.856E+10			
50	254000	12700000	2500	6.452E+10			
52	262000	13624000	2704	6.864E+10			
53	276000	14628000	2809	7.618E+10			
54	284000	15336000	2916	8.066E+10			
517	2563000	115608000	23211	5.767E+11			

TENSILE STRENGTH versus BRINELL HARDNESS

AISI Type P20 mold tool steel, Oil quenched, Hardness range (260-540) HB

$$\sigma = 545 \times HB - 11770$$

Brinell hardness *HB, X*	Tensile strength, psi *σ, Y*	*XY*	*X²*	*Y²*	Parameters of the regression line *slope*	*intercept*	Correlation coefficient
258	130000	33540000	66564	1.69E+10	545	-11770	0.998
286	144000	41184000	81796	2.074E+10			
327	168000	54936000	106929	2.822E+10			
353	180000	63540000	124609	3.24E+10			
381	192000	73152000	145161	3.686E+10			
409	207000	84663000	167281	4.285E+10			
432	224000	96768000	186624	5.018E+10			
455	242000	1.1E+08	207025	5.856E+10			
481	254000	1.22E+08	231361	6.452E+10			
512	262000	1.34E+08	262144	6.864E+10			
525	276000	1.45E+08	275625	7.618E+10			
543	284000	1.54E+08	294849	8.066E+10			
4962	2563000	1.113E+09	2149968	5.767E+11			

Relationship Between Hardness and Tensile Strength

AISI Type P20 Tool Steel, Hardness range (260-540) HB

Brinell hardness, HB (3000-kgf load)	Tensile strength, psi σ	Ratio σ/HB
258	130000	504
286	144000	503
327	168000	514
353	180000	510
381	192000	504
409	207000	506
432	224000	519
455	242000	532
481	254000	528
512	262000	512
525	276000	526
543	284000	523
4962	**2563000**	**6181**
average: 414	**average: 213583**	**average: 517**

Statistical Analysis of Data for the Tensile Strength (σ)-Brinell Hardness (HB) Ratio (σ/HB)

AISI Type P20 Tool Steel, Hardness range (260-540) HB

Observation	σ/HB	σ/HB-X	$(\sigma/HB-X)^2$
1	504	-16	256
2	503	-17	289
3	514	-6	36
4	510	-10	100
5	504	-16	256
6	506	-14	196
7	519	-1	1
8	532	12	144
9	528	8	64
10	512	-8	64
11	526	6	36
12	523	3	9
Average ratio: **X** = 520		**Sum of squares:** $\Sigma(\sigma/HB-X)^2 = 1451$	

Nomenclature	Symbol	Value
1. Number of observations	**n**	12
2. Number of degrees of freedom (d.f.)	**n - 1**	11
3. Sample mean (average σ/HB ratio)	**X**	520
4. Absolute error of the mean	α	3.3
5. Average σ/HB ratio range	$X \pm \alpha$	520 ± 3
6. Sample standard deviation	**S**	11.5
7. Coefficient of variation	**V**	2.2%
8. Relative error of the mean	ε	0.6%
9. *t* value at 95% two-sided confidence interval for d.f. = 11	$t_{.025}$	2.201
10. Population mean is greater than:	$X - \alpha_x t$	513
11. Population mean is less than:	$X + \alpha_x t$	527

Appendix VII

Hardness and Tensile Strength Relationship for Cast Irons

TENSILE STRENGTH versus BRINELL HARDNESS

ASTM A 48 class gray cast iron, Hardness range (160-300) HB

$$\sigma = 283 \times HB - 22970$$

Brinell hardness	Tensile strength, psi	XY	X^2	Y^2	Parameters of the regression line		Correlation coefficient
HB, X	***σ, Y***				***slope***	***intercept***	
156	22000	3432000	24336	484000000	283	-22970	0.998
174	26000	4524000	30276	676000000			
212	36500	7738000	44944	1.332E+09			
235	42500	9987500	55225	1.806E+09			
262	52500	13755000	68644	2.756E+09			
302	62500	18875000	91204	3.906E+09			
1341	242000	58311500	314629	1.096E+10			

COMPRESSIVE STRENGTH versus BRINELL HARDNESS

ASTM A 48 class gray cast iron, Hardness range (160-300) HB

$$\sigma_c = 725 \times HB - 29440$$

Brinell hardness	Compressive strength, psi	*XY*	X^2	Y^2	Parameters of the regression line		Correlation coefficient
HB, X	σ_c, ***Y***				***slope***	***intercept***	
156	83000	12948000	24336	6.889E+09	725	-29440	0.999
174	97000	16878000	30276	9.409E+09			
212	124000	26288000	44944	1.538E+10			
235	140000	32900000	55225	1.96E+10			
262	164000	42968000	68644	2.69E+10			
302	187500	56625000	91204	3.516E+10			
1341	795500	18860700	314629	1.133E+11			

Relationship Between Hardness and Tensile Strength

Ductile cast iron, Hardness range (140-390) HB

Specification	Grade or class	Brinell hardness, HB (3000-kgf load)	Tensile strength, psi σ	Ratio σ/HB
ASTM A 395	**60-40-18**	143	60000	420
SAE J434 (USA)	**D4512**	156	65000	417
SAE J434 (USA)	**D5506**	187	80000	428
ISO 1083 (International)	**500-7**	170	73000	429
ISO 1083 (International)	**600-3**	192	87000	453
ISO 1083 (International)	**700-2**	229	102000	445
ISO 1083 (International)	**800-2**	248	116000	468
ASTM A 897- 90	**125-80-10**	269	125000	465
Sulzer (Switzerland)	**GGG80**	250	116000	464
BCIRA (GB)	**950/6**	300	138000	460
BCIRA (GB)	**1050/3**	345	152000	441
BCIRA (GB)	**1200/1**	390	174000	44
		2879	**1288000**	**5336**
		average: 240	**average: 107333**	**average: 447**

Statistical Analysis of Data for the Tensile Strength (σ)-Brinell Hardness (HB) Ratio (σ/HB)

Ductile cast iron, Hardness range (140-390) HB

Observation	σ/HB	σ/HB-X	(σ/HB-X)^2
1	420	-30	900
2	417	-33	1089
3	428	-22	484
4	429	-21	441
5	453	3	9
6	445	-5	25
7	468	18	324
8	465	15	225
9	464	14	196
10	460	10	100
11	441	-9	81
12	446	-4	16
Average ratio: **X** = 450		**Sum of squares:** $\Sigma(\sigma/HB-X)^2 = 3890$	

Nomenclature	Symbol	Value
1. Number of observations	n	12
2. Number of degrees of freedom (d.f.)	n - 1	11
3. Sample mean (average σ/HB ratio)	X	450
4. Absolute error of the mean	α	5.4
5. Average σ/HB ratio range	X ± α	450 ± 5
6. Sample standard deviation	S	18.8
7. Coefficient of variation	V	4.2%
8. Relative error of the mean	ε	1.2%
9. *t* value at 95% two-sided confidence interval for d.f. = 11	$t_{.025}$	2.201
10. Population mean is greater than:	$X-\alpha \times t$	438
11. Population mean is less than:	$X+\alpha \times t$	462

TENSILE STRENGTH versus BRINELL HARDNESS

Ductile cast iron, Hardness range (140-390) HB

$$\sigma = 467 \times HB - 4720$$

Brinell hardness	Tensile strength, psi	XY	X^2	Y^2	Parameters of the regression line		Correlation coefficient
HB, X	***σ, Y***				***slope***	***intercept***	
143	60000	8580000	20449	3.6E+09	467	-4720	0.996
156	65000	10140000	24336	4.225E+09			
170	73000	12410000	28900	5.329E+09			
187	80000	14960000	34969	6.4E+09			
192	87000	16704000	36864	7.569E+09			
229	102000	23358000	52441	1.04E+10			
248	116000	28768000	61504	1.346E+10			
250	116000	29000000	62500	1.346E+10			
269	125000	33625000	72361	1.563E+10			
300	138000	41400000	90000	1.904E+10			
345	152000	52440000	119025	2.31E+10			
390	174000	67860000	152100	3.028E+10			
2879	1288000	339245000	755449	1.525E+11			

TENSILE STRENGTH versus BRINELL HARDNESS

Ductile cast iron: ASTM A 897-90 (302, 341, 388, 444) HB, GGG100 (310 HB), and GGG120 (330 HB)

$$\sigma = 536 \times HB - 7490$$

Brinell hardness	Tensile strength, psi	*XY*	X^2	Y^2	Parameters of the regression line		Correlation coefficient
HB, X	***σ, Y***				***slope***	***intercept***	
302	150000	45300000	91204	2.25E+10	536	-7490	0.995
310	160000	49600000	96100	2.56E+10			
330	174000	57420000	108900	3.028E+10			
341	175000	59675000	116281	3.06E+10			
388	200000	77600000	150544	4E+10			
444	230000	102120000	197136	5.29E+10			
2115	1089000	391715000	760165	2.02E+11			

Relationship Between Hardness and Tensile Strength

Malleable cast iron: ASTM A 220, ASTM A 602, SAE J158
Hardness range (150-270) HB

Class or grade	Brinell hardness, HB (3000-kgf load)	Tensile strength, psi σ	Ratio σ/HB
40010	149	60000	403
45006	156	65000	417
45008	156	65000	417
50005	179	70000	391
60004	197	80000	406
70003	217	85000	392
80002	241	95000	394
90001	269	105000	390
M4504	163	65000	399
M5003	187	75000	401
M5503	187	75000	401
M7002	229	90000	393
M8501	269	105000	390
	2599	**1035000**	**5194**
	average: 200	**average: 79615**	**average: 398**

Statistical Analysis of Data for the Tensile Strength (σ)-Brinell Hardness (HB) Ratio (σ/HB)

Malleable cast iron, Hardness range (150-270) HB

Observation	σ/HB	σ/HB-X	$(\sigma/HB-X)^2$
1	403	3	9
2	417	17	289
3	417	17	289
4	391	-9	81
5	406	6	36
6	392	-8	64
7	394	-6	36
8	390	-10	100
9	399	-1	1
10	401	1	1
11	401	1	1
12	393	-7	49
13	390	-10	100
Average ratio: X = 400		**Sum of squares:** $\Sigma(\sigma/HB-X)^2 = 1056$	

Nomenclature	Symbol	Value
1. Number of observations	**n**	13
2. Number of degrees of freedom (d.f.)	**n - 1**	12
3. Sample mean (average σ/HB ratio)	**X**	400
4. Absolute error of the mean	α	2.6
5. Average σ/HB ratio range	$X \pm \alpha$	400 ± 3
6. Sample standard deviation	**S**	9.4
7. Coefficient of variation	**V**	2.4%
8. Relative error of the mean	ε	0.7%
9. *t* value at 95% two-sided confidence interval for d.f. = 12	$t_{.025}$	2.179
10. Population mean is greater than:	$X-\alpha_x t$	394
11. Population mean is less than:	$X+\alpha_x t$	406

TENSILE STRENGTH versus BRINELL HARDNESS

Malleable cast iron: ASTM A 220, ASTM A 602, SAE J158,
Hardness range (150-270) HB

$$\sigma = 366 \times HB + 6420$$

Brinell hardness *HB, X*	Tensile strength, psi *σ, Y*	*XY*	X^2	Y^2	Parameters of the regression line *slope*	*intercept*	Correlation coefficient
149	60000	8940000	22201	3.6E+09	366	6420	0.998
156	65000	10140000	24336	4.225E+09			
156	65000	10140000	24336	4.225E+09			
163	65000	10595000	26569	4.225E+09			
179	70000	12530000	32041	4.9E+09			
187	75000	14025000	34969	5.625E+09			
187	75000	14025000	34969	5.625E+09			
197	80000	15760000	38809	6.4E+09			
217	85000	18445000	47089	7.225E+09			
229	90000	20610000	52441	8.1E+09			
241	95000	22895000	58081	9.025E+09			
269	105000	28245000	72361	1.103E+10			
269	105000	28245000	72361	1.103E+10			
2599	1035000	214595000	540563	8.523E+10			

Appendix VIII

Hardness and Tensile Strength Relationship for Titanium Alloys and Pure Titanium

Relationship Between Hardness and Tensile Strength

Alpha-beta Titanium Alloys
Hardness range (205-385) HB

Titanium alloy designation	Brinell hardness, HB (3000-kgf load)	Tensile strength, psi σ	Ratio σ/HB
Ti-5Al-2.5Sn-ELI	241	123000	510
Ti-5Al-2.5Sn-ELI	285	133000	467
Ti-3Al-2.5V	205	102000	498
Ti-6Al-4V	295	145000	492
Ti-6Al-4V	326	154000	472
Ti-8Al-1Mo-1V	300	154000	513
Ti-8Al-1Mo-1V	321	157000	489
Ti-6Al-6V-2Sn	326	154000	472
Ti-6Al-6V-2Sn	385	188000	488
	2684	**1310000**	**4401**
	average: 298	**average: 145556**	**average: 488**

Statistical Analysis of Data for the Tensile Strength (σ)-Brinell Hardness (HB) Ratio (σ/HB)

Alpha-beta Titanium Alloys, Hardness range (205-385) HB

Observation	σ/HB	σ/HB-X	$(\sigma/HB-X)^2$
1	510	20	400
2	467	-23	529
3	498	8	64
4	492	2	4
5	472	-18	324
6	513	23	529
7	489	-1	1
8	472	-18	324
9	488	-2	4
Average ratio: **X** = 490		**Sum of squares:** $\Sigma(\sigma/HB-X)^2 = 2179$	

Nomenclature	Symbol	Value
1. Number of observations	**n**	9
2. Number of degrees of freedom (d.f.)	**n - 1**	8
3. Sample mean (average σ/HB ratio)	**X**	490
4. Absolute error of the mean	α	5.5
5. Average σ/HB ratio range	$X \pm \alpha$	490 ± 6
6. Sample standard deviation	**S**	16.5
7. Coefficient of variation	**V**	3.4%
8. Relative error of the mean	ε	1.1%
9. *t* value at 95% two-sided confidence interval for d.f. = 8	$t_{.025}$	2.306
10. Population mean is greater than:	$X-\alpha_x t$	477
11. Population mean is less than:	$X+\alpha_x t$	503

TENSILE STRENGTH versus BRINELL HARDNESS

Alpha-beta Titanium Alloys, Hardness range (205-385) HB

$$\sigma = 457 \times HB + 9150$$

Brinell hardness	Tensile strength, psi	*XY*	X^2	Y^2	Parameters of the regression line		Correlation coefficient
HB, X	***σ, Y***				***slope***	***intercept***	
205	102000	20910000	42025	1.04E+10	457	9150	0.983
241	123000	29643000	58081	1.513E+10			
285	133000	37905000	81225	1.769E+10			
295	145000	42775000	87025	2.103E+10			
300	154000	46200000	90000	2.372E+10			
321	157000	50397000	103041	2.465E+10			
326	154000	50204000	106276	2.372E+10			
326	154000	50204000	106276	2.372E+10			
385	188000	72380000	148225	3.534E+10			
2684	1310000	400618000	822174	1.954E+11			

TENSILE STRENGTH versus ROCKWELL HARDNESS "C" SCALE

Alpha-beta Titanium Alloys, Hardness range (20-47) HRC

$$\sigma = 3300 \times HRC + 35230$$

Rockwell hardness	Tensile strength, psi	*XY*	X^2	Y^2	Parameters of the regression line		Correlation coefficient
HRC, X	***σ, Y***				***slope***	***intercept***	
19.6	102000	1999200	384.16	1.04E+10	3300	35230	0.983
28.0	123000	3444000	784	1.513E+10			
32.2	133000	4282600	1036.84	1.769E+10			
33.2	145000	4814000	1102.24	2.103E+10			
34.0	154000	5236000	1156	2.372E+10			
35.2	154000	5420800	1239.04	2.372E+10			
35.9	154000	5528600	1288.81	2.372E+10			
36.0	157000	5652000	1296	2.465E+10			
46.8	188000	8798400	2190.24	3.534E+10			
300.9	1310000	45175600	10477.33	1.954E+11			

Relationship Between Hardness and Tensile Strength

Alpha-beta Titanium Alloys
Hardness range (230-420) HK

Titanium alloy designation	Knoop hardness HK	Tensile strength, psi σ	Ratio σ/HK
Ti-5Al-2.5Sn-ELI	265	123000	464
Ti-5Al-2.5Sn-ELI	310	133000	429
Ti-3Al-2.5V	230	102000	443
Ti-6Al-4V	320	145000	453
Ti-6Al-4V	350	154000	440
Ti-8Al-1Mo-1V	325	154000	474
Ti-8Al-1Mo-1V	345	157000	455
Ti-6Al-6V-2Sn	350	154000	440
Ti-6Al-6V-2Sn	420	188000	448
	2915	**1310000**	**4046**
	average: 324	**average: 145556**	**average: 449**

Statistical Analysis of Data for the Tensile Strength (σ)-Knoop Hardness (HK) Ratio (σ/HK)

Alpha-beta Titanium Alloys, Hardness range (230-420) HK

Observation	σ/HK	σ/HK-X	$(\sigma/HK-X)^2$
1	464	14	196
2	429	-21	441
3	443	-7	49
4	453	3	9
5	440	-10	100
6	474	24	576
7	455	5	25
8	440	-10	100
9	448	-2	4
Average ratio: X = 450		**Sum of squares:** $\Sigma(\sigma/HK-X)^2 = 1500$	

Nomenclature	Symbol	Value
1. Number of observations	**n**	9
2. Number of degrees of freedom (d.f.)	**n - 1**	8
3. Sample mean (average σ/HK ratio)	**X**	450
4. Absolute error of the mean	α	4.6
5. Average σ/HK ratio range	$X \pm \alpha$	450 ± 5
6. Sample standard deviation	**S**	13.7
7. Coefficient of variation	**V**	3.0%
8. Relative error of the mean	ε	1.0%
9. *t* value at 95% two-sided confidence interval for d.f. = 8	$t_{.025}$	2.306
10. Population mean is greater than:	$X-\alpha_x\, t$	439
11. Population mean is less than:	$X+\alpha_x\, t$	461

TENSILE STRENGTH versus KNOOP HARDNESS

Alpha-beta Titanium Alloys, Hardness range (230-420) HK

$$\sigma = 440 \times HK + 2910$$

Knoop hardness *HK, X*	Tensile strength, psi *σ, Y*	*XY*	*X²*	*Y²*	Parameters of the regression line *slope*	Parameters of the regression line *intercept*	Correlation coefficient
230	102000	23460000	52900	1.04E+10	440	2910	0.984
265	123000	32595000	70225	1.513E+10			
310	133000	41230000	96100	1.769E+10			
320	145000	46400000	102400	2.103E+10			
325	154000	50050000	105625	2.372E+10			
345	157000	54165000	119025	2.465E+10			
350	154000	53900000	122500	2.372E+10			
350	154000	53900000	122500	2.372E+10			
420	188000	78960000	176400	3.534E+10			
2915	1310000	434660000	967675	1.954E+11			

TENSILE STRENGTH versus BRINELL HARDNESS

ASTM pure titanium grades 1 to 4, Hardness range (120-215) HB

$$\sigma = 510 \times HB - 6160$$

Brinell hardness HB, X	Tensile strength, psi σ, Y	XY	X^2	Y^2	Parameters of the regression line slope	intercept	Correlation coefficient
121	50000	6050000	14641	2.5E+09	510	-6160	0.961
144	67000	9648000	20736	4.489E+09			
153	73000	11169000	23409	5.329E+09			
153	79000	12087000	23409	6.241E+09			
195	88000	17160000	38025	7.744E+09			
190	96000	18240000	36100	9.216E+09			
215	101000	21715000	46225	1.02E+10			
1171	554000	96069000	202545	4.572E+10			

TENSILE STRENGTH versus KNOOP HARDNESS

ASTM pure titanium grades 1 to 4, Hardness range (140-240) HK

$$\sigma = 478 \times HK - 11730$$

Knoop hardness	Tensile strength, psi	XY	X^2	Y^2	Parameters of the regression line		Correlation coefficient
HK, X	***σ, Y***				***slope***	***intercept***	
140	50000	7000000	19600	2.5E+09	478	-11730	0.963
165	67000	11055000	27225	4.489E+09			
175	73000	12775000	30625	5.329E+09			
175	79000	13825000	30625	6.241E+09			
220	88000	19360000	48400	7.744E+09			
215	96000	20640000	46225	9.216E+09			
240	101000	24240000	57600	1.02E+10			
1330	554000	108895000	260300	4.572E+10			

Appendix IX

Hardness and Tensile Strength Relationship for Wrought Aluminum Alloys

Relationship Between Hardness and Tensile Strength

Aluminum alloys, 2xxx series (major alloying element-copper)
Hardness range (45-130) HB

Grade and temper*	Brinell hardness, HB (500-kgf load)	Tensile strength, psi σ	Ratio σ/HB
2011-T3	95	55000	579
2011-T8	100	59000	590
2014-0	45	27000	600
2014-T4	105	62000	590
2017-0	45	26000	578
2017-T4	105	62000	590
2024-0	47	27000	574
2024-T3	120	70000	583
2024-T4	120	68000	567
2024-T361	130	72000	554
2117-T4	70	43000	614
2618-T61	115	64000	557
	1097	**635000**	**6976**
	average: 91	**average: 52917**	**average: 580**

* Temper Designations:
0, Annealed to improve ductility.
T3, T361, Solution Heat Treated, Cold Worked, and Naturally Aged to a Substantially Stable Condition.
T4, Solution Heat Treated and Naturally Aged to a Substantially Stable Condition.
T6, T61, Solution Heat Treated and Artificially Aged.
T8, Solution Heat Treated, Cold Worked, and Artificially Aged.

Statistical Analysis of Data for the Tensile Strength (σ)-Brinell Hardness (HB) Ratio (σ/HB)

Aluminum alloys, 2xxx series, tempers: 0, T3, T4, T6, and T8
Hardness range (45-130) HB

Observation	σ/HB	σ/HB-X	$(\sigma/HB-X)^2$
1	579	-1	1
2	590	10	100
3	600	20	400
4	590	10	100
5	578	-2	4
6	590	10	100
7	574	-6	36
8	583	3	9
9	567	-13	169
10	554	-26	676
11	614	34	1156
12	557	-23	529
Average ratio: **X** = 580		**Sum of squares:** $\Sigma(\sigma/HB-X)^2 = 3280$	

Nomenclature	Symbol	Value
1. Number of observations	**n**	12
2. Number of degrees of freedom (d.f.)	**n - 1**	11
3. Sample mean (average σ/HB ratio)	**X**	580
4. Absolute error of the mean	α	5.0
5. Average σ/HB ratio range	$X \pm \alpha$	580 ± 5
6. Sample standard deviation	**S**	17.3
7. Coefficient of variation	**V**	3.0%
8. Relative error of the mean	ε	0.9%
9. *t* value at 95% two-sided confidence interval for d.f. = 11	$t_{.025}$	2.201
10. Population mean is greater than:	$X-\alpha_x t$	569
11. Population mean is less than:	$X+\alpha_x t$	591

TENSILE STRENGTH versus BRINELL HARDNESS

Aluminum alloys, 2xxx series (major alloying element - copper)
Tempers: 0, T3, T4, T6, and T8
Hardness range (45-130) HB measured at 500-kgf load

$$\sigma = 555 \times HB + 2150$$

Brinell hardness *HB, X*	Tensile strength, psi *σ, Y*	*XY*	*X²*	*Y²*	Parameters of the regression line *slope*	*intercept*	Correlation coefficient
45	26000	1170000	2025	676000000	555	2150	0.996
45	27000	1215000	2025	729000000			
47	27000	1269000	2209	729000000			
70	43000	3010000	4900	1.849E+09			
95	55000	5225000	9025	3.025E+09			
100	59000	5900000	10000	3.481E+09			
105	62000	6510000	11025	3.844E+09			
105	62000	6510000	11025	3.844E+09			
115	64000	7360000	13225	4.096E+09			
120	68000	8160000	14400	4.624E+09			
120	70000	8400000	14400	4.9E+09			
130	72000	9360000	16900	5.184E+09			
1097	635000	64089000	111159	3.698E+10			

Relationship Between Hardness and Tensile Strength

Aluminum alloys, 3xxx series (major alloying element-manganese)
Hardness range (28-77) HB

Grade and temper*	Brinell hardness, HB (500-kgf load)	Tensile strength, psi σ	Ratio σ/HB
3003-0	28	16000	571
3003-H12	35	19000	543
3003-H14	40	22000	550
3003-H16	47	26000	553
3003-H18	55	29000	527
3004-0	45	26000	578
3004-H32	52	31000	596
3004-H34	63	35000	556
3004-H36	70	38000	543
3004-H38	77	41000	532
	512	**283000**	**5549**
	average: 51	**average: 28300**	**average: 555**

* Temper Designations:
0, Annealed to improve ductility.
H1, Strain-Hardened to obtain the desired strength without supplementary thermal treatment.
H3, Strain-Hardened and Stabilized to improve ductility.
The digit following H1 and H3 indicates the degree of strain hardening.

Statistical Analysis of Data for the Tensile Strength (σ)-Brinell Hardness (HB) Ratio (σ/HB)

Aluminum alloys, 3xxx series, tempers: 0, H1, and H3
Hardness range (28-77) HB

Observation	σ/HB	σ/HB-X	$(\sigma/HB-X)^2$
1	571	16	256
2	543	-12	144
3	550	-5	25
4	553	-2	4
5	527	-28	784
6	578	23	529
7	596	41	1681
8	556	1	1
9	543	-12	144
10	532	-23	529
Average ratio: **X** = 555		**Sum of squares:** $\Sigma(\sigma/HB-X)^2 = 4097$	

Nomenclature	Symbol	Value
1. Number of observations	**n**	10
2. Number of degrees of freedom (d.f.)	**n - 1**	9
3. Sample mean (average σ/HB ratio)	**X**	555
4. Absolute error of the mean	α	6.7
5. Average σ/HB ratio range	$X \pm \alpha$	555 ± 7
6. Sample standard deviation	**S**	21.3
7. Coefficient of variation	**V**	3.8%
8. Relative error of the mean	ε	1.2%
9. *t* value at 95% two-sided confidence interval for d.f. = 9	$t_{.025}$	2.262
10. Population mean is greater than:	$X-\alpha_x t$	540
11. Population mean is less than:	$X+\alpha_x t$	570

TENSILE STRENGTH versus BRINELL HARDNESS

Aluminum alloys, 3xxx series (major alloying element - manganese)
Tempers: 0, H1, and H3
Hardness range (28-77) HB measured at 500-kgf load

$$\sigma = 522 \times HB + 1595$$

Brinell hardness	Tensile strength, psi	*XY*	X^2	Y^2	Parameters of the regression line		Correlation coefficient
HB, X	***σ, Y***				***slope***	***intercept***	
28	16000	448000	784	256000000	522	1595	0.992
35	19000	665000	1225	361000000			
40	22000	880000	1600	484000000			
45	26000	1170000	2025	676000000			
47	26000	1222000	2209	676000000			
52	31000	1612000	2704	961000000			
55	29000	1595000	3025	841000000			
63	35000	2205000	3969	1.225E+09			
70	38000	2660000	4900	1.444E+09			
77	41000	3157000	5929	1.681E+09			
512	283000	15614000	28370	8.605E+09			

Relationship Between Hardness and Tensile Strength

Aluminum alloys, 5xxx series (major alloying element-magnesium)
Hardness range (28-65) HB

Grade and temper*	Brinell hardness, HB (500-kgf load)	Tensile strength, psi σ	Ratio σ/HB
5005-0	28	18000	643
5050-0	36	21000	583
5052-0	47	28000	596
5056-0	65	42000	646
5154-0	58	35000	603
5254-0	58	35000	603
5454-0	62	36000	581
5457-0	32	19000	594
5652-0	47	28000	596
	433	**262000**	**5445**
	average: 48	**average: 29111**	**average: 605**

* Temper Designations:
0, Annealed to improve ductility.

Statistical Analysis of Data for the Tensile Strength (σ)-Brinell Hardness (HB) Ratio (σ/HB)

Aluminum alloys, 5xxx series, temper: 0 (annealed)
Hardness range (28-65) HB

Observation	σ/HB	σ/HB-X	$(\sigma/HB-X)^2$
1	643	38	1444
2	583	-22	484
3	596	-9	81
4	646	41	1681
5	603	-2	4
6	603	-2	4
7	581	-24	576
8	594	-11	121
9	596	-9	81
Average ratio: **X** = 605		**Sum of squares:** $\Sigma(\sigma/HB-X)^2 = 4476$	

Nomenclature	Symbol	Value
1. Number of observations	**n**	9
2. Number of degrees of freedom (d.f.)	**n - 1**	8
3. Sample mean (average σ/HB ratio)	**X**	605
4. Absolute error of the mean	α	7.9
5. Average σ/HB ratio range	$X \pm \alpha$	605 ± 8
6. Sample standard deviation	**S**	23.7
7. Coefficient of variation	**V**	3.9%
8. Relative error of the mean	ε	1.3%
9. *t* value at 95% two-sided confidence interval for d.f. = 8	$t_{.025}$	2.306
10. Population mean is greater than:	$X-\alpha \times t$	587
11. Population mean is less than:	$X+\alpha \times t$	623

TENSILE STRENGTH versus BRINELL HARDNESS

Aluminum alloys, 5xxx series (major alloying element - magnesium)
Temper: 0 (annealed)
Hardness range (28-65) HB measured at 500-kgf load

$$\sigma = 617 \times HB - 557$$

Brinell hardness	Tensile strength, psi	*XY*	X^2	Y^2	Parameters of the regression line		Correlation coefficient
HB, X	***σ, Y***				***slope***	***intercept***	
28	18000	504000	784	324000000	617	-557	0.990
32	19000	608000	1024	361000000			
36	21000	756000	1296	441000000			
47	28000	1316000	2209	784000000			
47	28000	1316000	2209	784000000			
58	35000	2030000	3364	1.225E+09			
58	35000	2030000	3364	1.225E+09			
62	36000	2232000	3844	1.296E+09			
65	42000	2730000	4225	1.764E+09			
433	262000	13522000	22319	8.204E+09			

Relationship Between Hardness and Tensile Strength

Aluminum alloys, 5xxx series (major alloying element-magnesium)
Hardness range (36-105) HB

Grade and temper*	Brinell hardness, HB (500-kgf load)	Tensile strength, psi σ	Ratio σ/HB
5005-H32	36	20000	556
5005-H34	41	23000	561
5005-H36	46	26000	565
5005-H38	51	29000	569
5050-H32	46	25000	543
5052-H32	60	33000	550
5052-H34	68	38000	559
5052-H36	73	40000	548
5052-H38	77	42000	545
5056-H18	105	63000	600
5056-H38	100	60000	600
5154-H32	67	39000	582
5154-H34	73	42000	575
5154-H36	78	45000	577
5154-H38	80	48000	600
5154-H112	63	35000	556
5252-H38	75	41000	547
5254-H32	67	39000	582
5254-H34	73	42000	575
5254-H36	78	45000	577
5254-H38	80	48000	600
5254-H112	63	35000	556
5454-H32	73	40000	548
5454-H34	81	44000	543
5454-H111	70	38000	543
5454-H112	62	36000	581
5456-H321	90	51000	567
5457-H25	48	26000	542
5457-H38	55	30000	545
5652-H32	60	33000	550
5652-H34	68	38000	559
5652-H36	73	40000	548
5652-H38	77	42000	545
5657-H25	40	23000	575
5657-H38	50	28000	560
	2347	**1327000**	**19729**
	average: 67	**average: 37914**	**average: 565**

* Temper Designations:
H1, Strain-Hardened to obtain the desired strength without supplementary thermal treatment.
H2, Strain-Hardened and Partially Annealed.
H3, Strain-Hardened and Stabilized to improve ductility.
The digit following H1, H2, and H3 indicates the degree of strain hardening.

Statistical Analysis of Data for the Tensile Strength (σ)-Brinell Hardness (HB) Ratio (σ/HB)

Aluminum alloys, 5xxx series, temper: H (strain-hardened)
Hardness range (36-105) HB

Observation	σ/HB	σ/HB-X	(σ/HB-X)^2
1	556	-9	81
2	561	-4	16
3	565	0	0
4	569	4	16
5	543	-22	484
6	550	-15	225
7	559	-6	36
8	548	-17	289
9	545	-20	400
10	600	35	1225
11	600	35	1225
12	582	17	289
13	575	10	100
14	577	12	144
15	600	35	1225
16	556	-9	81
17	547	-18	324
18	582	17	289
19	575	10	100
20	577	12	144
21	600	35	1225
22	556	-9	81
23	548	-17	289
24	543	-22	484
25	543	-22	484
26	581	16	256
27	567	2	4
28	542	-23	529
29	545	-20	400
30	550	-15	225
31	559	-6	36
32	548	-17	289
33	545	-20	400
34	575	10	100
35	560	-5	25
Average ratio: **X** = 565		**Sum of squares:** **Σ(σ/HB-X)^2** = 11520	

(continued on next page)

Statistical Analysis of Data for the Tensile Strength (σ)-Brinell Hardness (HB) Ratio (σ/HB)

Aluminum alloys, 5xxx series, temper: H (strain-hardened)
Hardness range (36-105) HB
(continued)

Nomenclature	Symbol	Value
1. Number of observations	**n**	35
2. Number of degrees of freedom (d.f.)	**n - 1**	34
3. Sample mean (average σ/HB ratio)	**X**	565
4. Absolute error of the mean	α	3.1
5. Average σ/HB ratio range	$X \pm \alpha$	565 ± 3
6. Sample standard deviation	**S**	18.4
7. Coefficient of variation	**V**	3.3%
8. Relative error of the mean	ε	0.5%
9. *t* value at 95% two-sided confidence interval for d.f. = 34	$t_{.025}$	2.042
10. Population mean is greater than:	$X - \alpha_x t$	559
11. Population mean is less than:	$X + \alpha_x t$	571

TENSILE STRENGTH versus BRINELL HARDNESS

Aluminum alloys, 5xxx series (major alloying element-magnesium)
Temper: H (strain-hardened). Hardness range (36-105) HB measured at 500-kgf load

$$\sigma = 604 \times HB - 2580$$

Brinell hardness	Tensile strength, psi	XY	X^2	Y^2	Parameters of the regression line		Correlation coefficient
HB, X	σ, Y				slope	intercept	
36	20000	720000	1296	400000000	604	-2580	0.991
40	23000	920000	1600	529000000			
41	23000	943000	1681	529000000			
46	25000	1150000	2116	625000000			
46	26000	1196000	2116	676000000			
48	26000	1248000	2304	676000000			
50	28000	1400000	2500	784000000			
51	29000	1479000	2601	841000000			
55	30000	1650000	3025	900000000			
60	33000	1980000	3600	1.089E+09			
60	33000	1980000	3600	1.089E+09			
62	36000	2232000	3844	1.296E+09			
63	35000	2205000	3969	1.225E+09			
63	35000	2205000	3969	1.225E+09			
67	39000	2613000	4489	1.521E+09			
67	39000	2613000	4489	1.521E+09			
68	38000	2584000	4624	1.444E+09			
68	38000	2584000	4624	1.444E+09			
70	38000	2660000	4900	1.444E+09			
73	40000	2920000	5329	1.6E+09			
73	40000	2920000	5329	1.6E+09			
73	40000	2920000	5329	1.6E+09			
73	42000	3066000	5329	1.764E+09			
73	42000	3066000	5329	1.764E+09			
75	41000	3075000	5625	1.681E+09			
77	42000	3234000	5929	1.764E+09			
77	42000	3234000	5929	1.764E+09			
78	45000	3510000	6084	2.025E+09			
78	45000	3510000	6084	2.025E+09			
80	48000	3840000	6400	2.304E+09			
80	48000	3840000	6400	2.304E+09			
81	44000	3564000	6561	1.936E+09			
90	51000	4590000	8100	2.601E+09			
100	60000	6000000	10000	3.6E+09			
105	63000	6615000	11025	3.969E+09			
2347	1327000	94266000	166129	5.356E+10			

Relationship Between Hardness and Tensile Strength

Aluminum alloys, 6xxx series (major alloying elements-magnesium and silicon)
Hardness range (25-90) HB

Grade and temper*	Brinell hardness, HB (500-kgf load)	Tensile strength, psi σ	Ratio σ/HB
6061-T4	65	35000	538
6063-0	25	13000	520
6063-T1	42	22000	524
6066-0	43	22000	512
6066-T4	90	52000	578
6463-T1	42	22000	524
	307	**166000**	**3196**
	average: 51	**average: 27667**	**average: 540**

* Temper Designations:
0, Annealed to improve ductility.
T1, Cooled From an Elevated-Temperature Shaping Process
and Naturally Aged to a Substantially Stable Condition.
T4, Solution Heat Treated and Naturally Aged to a Substantially Stable Condition.

Statistical Analysis of Data for the Tensile Strength (σ)-Brinell Hardness (HB) Ratio (σ/HB)

Aluminum alloys, 6xxx series, tempers: 0, T1, and T4
Hardness range (25-90) HB

Observation	σ/HB	σ/HB-X	$(\sigma/HB-X)^2$
1	538	-2	4
2	520	-20	400
3	524	-16	256
4	512	-28	784
5	578	38	1444
6	524	-16	256
Average ratio: X = 540		**Sum of squares:** $\Sigma(\sigma/HB-X)^2 = 3144$	

Nomenclature	Symbol	Value
1. Number of observations	n	6
2. Number of degrees of freedom (d.f.)	n - 1	5
3. Sample mean (average σ/HB ratio)	X	540
4. Absolute error of the mean	α	10.2
5. Average σ/HB ratio range	X ± α	540 ± 10
6. Sample standard deviation	S	25.1
7. Coefficient of variation	V	4.6%
8. Relative error of the mean	ε	1.9%
9. *t* value at 95% two-sided confidence interval for d.f. = 5	$t_{.025}$	2.571
10. Population mean is greater than:	$X-\alpha_x t$	514
11. Population mean is less than:	$X+\alpha_x t$	566

TENSILE STRENGTH versus BRINELL HARDNESS

Aluminum alloys, 6xxx series (major alloying elements - magnesium and silicon)
Tempers: 0, T1, and T4
Hardness range (25-90) HB measured at 500-kgf load

$$\sigma = 603 \times HB - 3210$$

Brinell hardness	Tensile strength, psi	XY	X^2	Y^2	Parameters of the regression line		Correlation coefficient
HB, X	σ, Y				slope	intercept	
25	13000	325000	625	169000000	603	-3210	0.998
42	22000	924000	1764	484000000			
42	22000	924000	1764	484000000			
43	22000	946000	1849	484000000			
65	35000	2275000	4225	1.225E+09			
90	52000	4680000	8100	2.704E+09			
307	166000	10074000	18327	5.55E+09			

Relationship Between Hardness and Tensile Strength

Aluminum alloys, 6xxx series (major alloying elements-magnesium and silicon)
Hardness range (60-120) HB

Grade and temper*	Brinell hardness, HB (500-kgf load)	Tensile strength, psi σ	Ratio σ/HB
6061-T6	95	45000	474
6063-T5	60	27000	450
6063-T6	73	35000	479
6063-T83	82	37000	451
6063-T831	70	30000	429
6063-T832	95	42000	442
6066-T6	120	57000	475
6101-T6	71	32000	451
6351-T6	95	45000	474
6463-T5	60	27000	450
6463-T6	74	35000	473
	895	**412000**	**5048**
	average: 81	**average: 37455**	**average: 460**

* Temper Designations:
T5, Cooled From an Elevated-Temperature Shaping Process and Artificially Aged.
T6, Solution Heat Treated and Artificially Aged.
T8, Solution Heat Treated, Cold Worked, and Artificially Aged.

Statistical Analysis of Data for the Tensile Strength (σ)-Brinell Hardness (HB) Ratio (σ/HB)

Aluminum alloys, 6xxx series, temper: T5, T6, and T8
Hardness range (60-120) HB

Observation	σ/HB	σ/HB-X	$(\sigma/HB-X)^2$
1	474	14	196
2	450	-10	100
3	479	19	361
4	451	-9	81
5	429	-31	961
6	442	-18	324
7	475	15	225
8	451	-9	81
9	474	14	196
10	450	-10	100
11	473	13	169
Average ratio: **X** = 460		**Sum of squares:** $\Sigma(\sigma/HB-X)^2 = 2794$	

Nomenclature	Symbol	Value
1. Number of observations	**n**	11
2. Number of degrees of freedom (d.f.)	**n - 1**	10
3. Sample mean (average σ/HB ratio)	**X**	460
4. Absolute error of the mean	α	5.0
5. Average σ/HB ratio range	$X \pm \alpha$	460 ± 5
6. Sample standard deviation	**S**	16.7
7. Coefficient of variation	**V**	3.6%
8. Relative error of the mean	ε	1.1%
9. *t* value at 95% two-sided confidence interval for d.f. = 10	$t_{.025}$	2.228
10. Population mean is greater than:	$X-\alpha_x t$	449
11. Population mean is less than:	$X+\alpha_x t$	471

TENSILE STRENGTH versus BRINELL HARDNESS

Aluminum alloys, 6xxx series (major alloying elements-magnesium and silicon)
Tempers: T5, T6, and T8
Hardness range (60-120) HB measured at 500-kgf load

$$\sigma = 496 \times HB - 2910$$

Brinell hardness	Tensile strength, psi	XY	X^2	Y^2	Parameters of the regression line		Correlation coefficient
HB, X	***σ, Y***				***slope***	***intercept***	
60	27000	1620000	3600	729000000	496	-2910	0.991
60	27000	1620000	3600	729000000			
70	30000	2100000	4900	900000000			
71	32000	2272000	5041	1.024E+09			
73	35000	2555000	5329	1.225E+09			
74	35000	2590000	5476	1.225E+09			
82	37000	3034000	6724	1.369E+09			
95	42000	3990000	9025	1.764E+09			
95	45000	4275000	9025	2.025E+09			
95	45000	4275000	9025	2.025E+09			
120	57000	6840000	14400	3.249E+09			
895	412000	35171000	76145	1.626E+10			

Relationship Between Hardness and Tensile Strength

Aluminum alloys, 7xxx series (major alloying element-zinc)
Hardness range (60-150) HB

Grade and temper*	Brinell hardness, HB (500-kgf load)	Tensile strength, psi σ	Ratio σ/HB
7049-T73	135	75000	556
7049-T7352	135	75000	556
7075-0	60	33000	550
7075-T6	150	83000	553
	480	**266000**	**2215**
	average: 120	**average: 66500**	**average: 554**

* Temper Designations:
0, Annealed to improve ductility.
T6, Solution Heat Treated and Artificially Aged.
T73, T7352, Solution Heat Treated and Overaged or Stabilized.

Statistical Analysis of Data for the Tensile Strength (σ)-Brinell Hardness (HB) Ratio (σ/HB)

Aluminum alloys, 7xxx series, tempers: 0, T6, and T7
Hardness range (60-150) HB

Observation	σ/HB	σ/HB-X	$(\sigma/HB-X)^2$
1	556	2	4
2	556	2	4
3	550	-4	16
4	553	-1	1
Average ratio: **X** = 554		**Sum of squares:** $\Sigma(\sigma/HB-X)^2 = 25$	

Nomenclature	Symbol	Value
1. Number of observations	**n**	4
2. Number of degrees of freedom (d.f.)	**n - 1**	3
3. Sample mean (average σ/HB ratio)	**X**	554
4. Absolute error of the mean	α	1.45
5. Average σ/HB ratio range	$X \pm \alpha$	554 ± 1
6. Sample standard deviation	**S**	2.9
7. Coefficient of variation	**V**	0.5%
8. Relative error of the mean	ε	0.3%
9. *t* value at 95% two-sided confidence interval for d.f. = 3	$t_{.025}$	3.182
10. Population mean is greater than:	$X-\alpha_x t$	549
11. Population mean is less than:	$X+\alpha_x t$	559

TENSILE STRENGTH versus BRINELL HARDNESS

Aluminum alloys, 7xxx series (major alloying element - zinc)
Tempers: 0, T6, and T7
Hardness range (60-150) HB measured at 500-kgf load

$$\sigma = 558 \times HB - 409$$

Brinell hardness	Tensile strength, psi	*XY*	X^2	Y^2	Parameters of the regression line		Correlation coefficient
HB, X	***σ, Y***				***slope***	***intercept***	
60	33000	1980000	3600	1.089E+09	558	-409	0.99997
135	75000	10125000	18225	5.625E+09			
135	75000	10125000	18225	5.625E+09			
150	83000	12450000	22500	6.889E+09			
480	266000	34680000	62550	1.923E+10			

Appendix X

Hardness and Tensile Strength Relationship for Cast Aluminum Alloys

TENSILE STRENGTH versus BRINELL HARDNESS

Aluminum alloy 204.0 (4.2-5.0% Cu)
Tempers: T4 and T6
Hardness range (105-125) HB measured at 500-kgf load

$$\sigma = 200 \times HB + 36000$$

Brinell hardness	Tensile strength, psi	*XY*	X^2	Y^2	Parameters of the regression line		Correlation coefficient
HB, X	***σ, Y***				***slope***	***intercept***	
105	57000	5985000	11025	3.249E+09	200	36000	1.000
110	58000	6380000	12100	3.364E+09			
115	59000	6785000	13225	3.481E+09			
125	61000	7625000	15625	3.721E+09			
455	235000	26775000	51975	1.382E+10			

TENSILE STRENGTH versus BRINELL HARDNESS

Aluminum alloy 242.0 (3.5-4.5% Cu)
Tempers: T21, T77, T571, and T61
Hardness range (70-110) HB measured at 500-kgf load

$$\sigma = 446 \times HB - 4465$$

Brinell hardness	Tensile strength, psi	*XY*	X^2	Y^2	Parameters of the regression line		Correlation coefficient
HB, X	*σ, Y*				*slope*	*intercept*	
70	27000	1890000	4900	729000000	446	-4465	0.972
75	30000	2250000	5625	900000000			
85	32000	2720000	7225	1.024E+09			
105	40000	4200000	11025	1.6E+09			
110	47000	5170000	12100	2.209E+09			
445	176000	16230000	40875	6.462E+09			

TENSILE STRENGTH versus BRINELL HARDNESS

Aluminum alloy 355.0 (4.5-5.5% Si, 1.0-1.5% Cu)
Tempers: T51, T7, T71, T6, and T61
Hardness range (65-105) HB measured at 500-kgf load

$$\sigma = 448 \times HB - 645$$

Brinell hardness	Tensile strength, psi	XY	X^2	Y^2	Parameters of the regression line		Correlation coefficient
HB, X	σ, Y				*slope*	*intercept*	
65	28000	1820000	4225	784000000	448	-645	0.936
75	30000	2250000	5625	900000000			
75	35000	2625000	5625	1.225E+09			
80	35000	2800000	6400	1.225E+09			
85	36000	3060000	7225	1.296E+09			
85	38000	3230000	7225	1.444E+09			
85	40000	3400000	7225	1.6E+09			
90	39000	3510000	8100	1.521E+09			
90	42000	3780000	8100	1.764E+09			
105	45000	4725000	11025	2.025E+09			
835	368000	31200000	70775	1.378E+10			

TENSILE STRENGTH versus BRINELL HARDNESS

Aluminum alloy 356.0 (6.5-7.5% Si, 0.25% Cu)
Tempers: T51, T71, T7, and T6
Hardness range (60-80) HB measured at 500-kgf load

$$\sigma = 556 \times HB - 6780$$

Brinell hardness	Tensile strength, psi	*XY*	X^2	Y^2	Parameters of the regression line		Correlation coefficient
HB, X	***σ, Y***				***slope***	***intercept***	
60	25000	1500000	3600	625000000	556	-6780	0.970
60	28000	1680000	3600	784000000			
70	32000	2240000	4900	1.024E+09			
70	33000	2310000	4900	1.089E+09			
75	34000	2550000	5625	1.156E+09			
80	38000	3040000	6400	1.444E+09			
415	190000	13320000	29025	6.122E+09			

TENSILE STRENGTH versus BRINELL HARDNESS

Aluminum alloy 390.0 (16.0-18.0% Si, 4.0-5.0% Cu)
Tempers: F (As-Fabricated), T5, T6, and T7
Hardness range (100-150) HB measured at 500-kgf load

$$\sigma = 511 \times HB - 24820$$

Brinell hardness *HB, X*	Tensile strength, psi *σ, Y*	*XY*	X^2	Y^2	Parameters of the regression line *slope*	*intercept*	Correlation coefficient
100	26000	2600000	10000	676000000	511	-24820	0.959
110	29000	3190000	12100	841000000			
110	30000	3300000	12100	900000000			
110	30000	3300000	12100	900000000			
115	36000	4140000	13225	1.296E+09			
120	38000	4560000	14400	1.444E+09			
125	40000	5000000	15625	1.6E+09			
125	43000	5375000	15625	1.849E+09			
145	45000	6525000	21025	2.025E+09			
150	53000	7950000	22500	2.809E+09			
1210	370000	45940000	148700	1.434E+10			

TENSILE STRENGTH versus BRINELL HARDNESS

Aluminum alloy 771.0 (6.5-7.5% Zn)
Tempers: T52, T6, T5, and T71
Hardness range (85-120) HB measured at 500-kgf load

$$\sigma = 292 \times HB + 13150$$

Brinell hardness	Tensile strength, psi	*XY*	X^2	Y^2	Parameters of the regression line		Correlation coefficient
HB, X	***σ, Y***				***slope***	***intercept***	
85	36000	3060000	7225	1.296E+09	292	13150	0.923
90	42000	3780000	8100	1.764E+09			
100	42000	4200000	10000	1.764E+09			
120	48000	5760000	14400	2.3-4E+09			
395	168000	16800000	39725	7.128E+09			

Appendix XI

Hardness and Tensile Strength Relationship for Wrought Copper Alloys

TENSILE STRENGTH versus ROCKWELL HARDNESS "B" SCALE

Wrought copper alloys, C2xxxx series (alloying element - zinc)
Hardness range (55-82) HRB

$$\sigma = 933 \times HRB - 1990$$

Rockwell hardness	Tensile strength, psi				Parameters of the regression line		Correlation coefficient
HRB, X	***σ, Y***	***XY***	***X^2***	***Y^2***	***slope***	***intercept***	
55	50000	2750000	3025	2.5E+09	933	-1990	0.980
58	52000	3016000	3364	2.70E+09			
60	55000	3300000	3600	3.025E+09			
69	60000	4140000	4761	3.6E+09			
77	70000	5390000	5929	4.9E+09			
80	70000	5600000	6400	4.9E+09			
82	78000	6396000	6724	6.084E+09			
481	435000	30592000	33803	2.771E+10			

TENSILE STRENGTH versus BRINELL HARDNESS

Wrought copper alloys, C2xxxx series (alloying element - zinc)
Hardness range (100-160) HB measured at 3000-kgf load

$$\sigma = 453 \times HB + 5150$$

Brinell hardness	Tensile strength, psi	XY	X^2	Y^2	Parameters of the regression line		Correlation coefficient
HB, X	***σ, Y***				***slope***	***intercept***	
100	50000	5000000	10000	2.5E+09	453	5150	0.987
104	52000	5408000	10816	2.704E+09			
107	55000	5885000	11449	3.025E+09			
123	60000	7380000	15129	3.6E+09			
141	70000	9870000	19881	4.9E+09			
150	70000	10500000	22500	4.9E+09			
156	78000	12168000	24336	6.084E+09			
881	435000	56211000	114111	2.771E+10			

TENSILE STRENGTH versus ROCKWELL HARDNESS "B" SCALE

Wrought copper alloys, C3xxxx series (alloying elements-zinc and lead)
Hardness range (60-85) HRB

$$\sigma = 796 \times HRB + 8370$$

Rockwell hardness	Tensile strength, psi	*XY*	X^2	Y^2	Parameters of the regression line		Correlation coefficient
HRB, X	***σ, Y***				***slope***	***intercept***	
60	55000	3300000	3600	3.025E+09	796	8370	0.956
60	58000	3480000	3600	3.364E+09			
62	56000	3472000	3844	3.136E+09			
65	58000	3770000	4225	3.364E+09			
68	63000	4284000	4624	3.996E+09			
70	63000	4410000	4900	3.996E+09			
70	65000	4550000	4900	4.225E+09			
70	65000	4550000	4900	4.225E+09			
70	65000	4550000	4900	4.225E+09			
72	65000	4680000	5184	4.225E+09			
72	67000	4824000	5184	4.489E+09			
72	68000	4896000	5184	4.624E+09			
75	70000	5250000	5625	4.9E+09			
80	68000	5440000	6400	4.624E+09			
80	70000	5600000	6400	4.9E+09			
85	75000	6375000	7225	5.625E+09			
85	75000	6375000	7225	5.625E+09			
85	80000	6800000	7225	6.4E+09			
1301	1186000	86606000	95145	7.891E+10			

TENSILE STRENGTH versus BRINELL HARDNESS

Wrought copper alloys, C3xxxx series (alloying elements-zinc and lead)
Hardness range (110-165) HB measured at 3000-kgf load

$$\sigma = 332 \times HB + 21900$$

Brinell hardness *HB, X*	Tensile strength, psi *σ, Y*	*XY*	X^2	Y^2	Parameters of the regression line *slope*	*intercept*	Correlation coefficient
107	55000	5885000	11449	3.025E+09	332	21900	0.947
107	58000	6206000	11449	3.364E+09			
110	56000	6160000	12100	3.136E+09			
116	58000	6728000	13456	3.364E+09			
121	63000	7623000	14641	3.969E+09			
125	63000	7875000	15625	3.969E+09			
125	65000	8125000	15625	4.225E+09			
125	65000	8125000	15625	4.225E+09			
125	65000	8125000	15625	4.225E+09			
130	65000	8450000	16900	4.225E+09			
130	67000	8710000	16900	4.489E+09			
130	68000	8840000	16900	4.624E+09			
137	70000	9590000	18769	4.9E+09			
150	68000	10200000	22500	4.624E+09			
150	70000	10500000	22500	4.9E+09			
165	75000	12375000	27225	5.625E+09			
165	75000	12375000	27225	5.625E+09			
165	80000	13200000	27225	6.4E+09			
2383	1186000	159092000	321739	7.891E+10			

TENSILE STRENGTH versus ROCKWELL HARDNESS "B" SCALE

Wrought copper alloys, C46xxx series (alloying elements-zinc and tin)
Hardness range (55-95) HRB

$$\sigma = 680 \times HRB + 19620$$

Rockwell hardness	Tensile strength, psi	*XY*	X^2	Y^2	Parameters of the regression line		Correlation coefficient
HRB, X	***σ, Y***				***slope***	***intercept***	
55	55000	3025000	3025	3.025E+09	680	19620	0.966
55	56000	3080000	3025	3.136E+09			
55	57000	3135000	3025	3.249E+09			
56	58000	3248000	3136	3.364E+09			
56	58000	3248000	3136	3.364E+09			
58	60000	3480000	3364	3.6E+09			
60	62000	3720000	3600	3.844E+09			
60	63000	3780000	3600	3.969E+09			
60	63000	3780000	3600	3.969E+09			
75	67000	5025000	5625	4.489E+09			
78	69000	5382000	6084	4.761E+09			
80	70000	5600000	6400	4.9E+09			
82	75000	6150000	6724	5.625E+09			
85	80000	6800000	7225	6.4E+09			
95	88000	8360000	9025	7.744E+09			
1010	981000	67813000	70594	6.544E+10			

TENSILE STRENGTH versus BRINELL HARDNESS

Wrought copper alloys, C46xxx series (alloying elements-zinc and tin)
Hardness range (100-210) HB measured at 3000-kgf load

$$\sigma = 282 \times HB + 29890$$

Brinell hardness	Tensile strength, psi	*XY*	X^2	Y^2	Parameters of the regression line		Correlation coefficient
HB, X	***σ, Y***				***slope***	***intercept***	
100	55000	5500000	10000	3.025E+09	282	29890	0.976
100	56000	5600000	10000	3.136E+09			
100	57000	5700000	10000	3.249E+09			
101	58000	5858000	10201	3.364E+09			
101	58000	5858000	10201	3.364E+09			
104	60000	6240000	10816	3.6E+09			
107	62000	6634000	11449	3.844E+09			
107	63000	6741000	11449	3.969E+09			
107	63000	6741000	11449	3.969E+09			
137	67000	9179000	18769	4.489E+09			
144	69000	9936000	20736	4.761E+09			
150	70000	10500000	22500	4.9E+09			
156	75000	11700000	24336	5.625E+09			
165	80000	13200000	27225	6.4E+09			
210	88000	18480000	44100	7.744E+09			
1889	981000	127867000	253231	6.544E+10			

TENSILE STRENGTH versus ROCKWELL HARDNESS "B" SCALE

Wrought copper alloys, C48xxx series (alloying elements-zinc, tin, and lead)
Hardness range (55-82) HRB

$$\sigma = 546 \times HRB + 27820$$

Rockwell hardness	Tensile strength, psi	*XY*	X^2	Y^2	Parameters of the regression line		Correlation coefficient
HRB, X	***σ, Y***				***slope***	***intercept***	
55	56000	3080000	3025	3.136E+09	546	27820	0.953
55	57000	3135000	3025	3.249E+09			
55	57000	3135000	3025	3.249E+09			
60	62000	3720000	3600	3.844E+09			
60	63000	3780000	3600	3.969E+09			
60	63000	3780000	3600	3.969E+09			
75	66000	4950000	5625	4.356E+09			
75	67000	5025000	5625	4.489E+09			
78	69000	5382000	6084	4.761E+09			
78	69000	5382000	6084	4.761E+09			
78	70000	5460000	6084	4.9E+09			
82	75000	6150000	6724	5.625E+09			
82	75000	6150000	6724	5.625E+09			
893	849000	59129000	62825	5.593E+10			

TENSILE STRENGTH versus BRINELL HARDNESS

Wrought copper alloys, C48xxx series (alloying elements-zinc, tin, and lead)
Hardness range (100-160) HB measured at 3000-kgf load

$$\sigma = 269 \times HB + 31350$$

Brinell hardness	Tensile strength, psi	*XY*	X^2	Y^2	Parameters of the regression line		Correlation coefficient
HB, X	***σ, Y***				***slope***	***intercept***	
100	56000	5600000	10000	3.136E+09	269	31350	0.955
100	57000	5700000	10000	3.249E+09			
100	57000	5700000	10000	3.249E+09			
107	62000	6634000	11449	3.844E+09			
107	63000	6741000	11449	3.969E+09			
107	63000	6741000	11449	3.969E+09			
137	66000	9042000	18769	4.356E+09			
137	67000	9179000	18769	4.489E+09			
144	69000	9936000	20736	4.761E+09			
144	69000	9936000	20736	4.761E+09			
144	70000	10080000	20736	4.9E+09			
156	75000	11700000	24336	5.625E+09			
156	75000	11700000	24336	5.625E+09			
1639	849000	108689000	212765	5.593E+10			

TENSILE STRENGTH versus ROCKWELL HARDNESS "B" SCALE

Wrought copper alloys, C5xxxx series (major alloying element - tin)
Hardness range (70-85) HRB

$$\sigma = 1259 \times HRB - 28520$$

Rockwell hardness	Tensile strength, psi	*XY*	X^2	Y^2	Parameters of the regression line		Correlation coefficient
HRB, X	***σ, Y***				***slope***	***intercept***	
70	60000	4200000	4900	3.6E+09	1259	-28520	0.939
78	70000	5460000	6084	4.9E+09			
80	68000	5440000	6400	4.624E+09			
80	75000	6000000	6400	5.625E+09			
83	75000	6225000	6889	5.625E+09			
85	80000	6800000	7225	6.4E+09			
476	428000	34125000	37898	3.077E+10			

TENSILE STRENGTH versus BRINELL HARDNESS

Wrought copper alloys, C5xxxx series (major alloying element - tin)
Hardness range (125-165) HB measured at 3000-kgf load

$$\sigma = 474 \times HB + 810$$

Brinell hardness	Tensile strength, psi	*XY*	X^2	Y^2	Parameters of the regression line		Correlation coefficient
HB, X	***σ, Y***				***slope***	***intercept***	
125	60000	7500000	15625	3.6E+09	474	810	0.941
144	70000	10080000	20736	4.9E+09			
150	68000	10200000	22500	4.624E+09			
150	75000	11250000	22500	5.625E+09			
159	75000	11925000	25281	5.625E+09			
165	80000	13200000	27225	6.4E+09			
893	428000	64155000	133867	3.077E+10			

TENSILE STRENGTH versus BRINELL HARDNESS

Wrought copper alloys, C6xxxx series (major alloying element - aluminum)
Hardness range (155-200) HB measured at 3000-kgf load

$$\sigma = 583 \times HB - 4570$$

Brinell hardness *HB, X*	Tensile strength, psi *σ, Y*	*XY*	X^2	Y^2	Parameters of the regression line *slope*	*intercept*	Correlation coefficient
155	84600	13113000	24025	7.157E+09	583	-4570	0.991
157	85500	13423500	24649	7.31E+09			
161	90000	14490000	25921	8.1E+09			
165	94500	15592500	27225	8.93E+09			
200	111100	22220000	40000	1.234E+10			
200	112500	22500000	40000	1.266E+10			
1038	578200	101339000	181820	5.65E+10			

TENSILE STRENGTH versus ROCKWELL HARDNESS "B" SCALE

Wrought copper alloys, C65xxx series (alloying element - silicon)
Hardness range (60-95) HRB

$$\sigma = 1236 \times HRB - 19010$$

Rockwell hardness	Tensile strength, psi	*XY*	X^2	Y^2	Parameters of the regression line		Correlation coefficient
HRB, X	***σ, Y***				***slope***	***intercept***	
60	58000	3480000	3600	3.364E+09	1236	-19010	0.938
85	78000	6630000	7225	6.08E+09			
90	90000	8100000	8100	8.1E+09			
90	92000	8280000	8100	8.464E+09			
92	93000	8556000	8464	8.649E+09			
95	108000	10260000	9025	1.166E+10			
512	519000	45306000	44514	4.633E+10			

TENSILE STRENGTH versus BRINELL HARDNESS

Wrought copper alloys, C65xxx series (alloying element - silicon)
Hardness range (110-210) HB measured at 3000-kgf load

$$\sigma = 458 \times HB + 6670$$

Brinell hardness	Tensile strength, psi	*XY*	X^2	Y^2	Parameters of the regression line		Correlation coefficient
HB, X	***σ, Y***				***slope***	***intercept***	
107	58000	6206000	11449	3.364E+09	458	6670	0.978
165	78000	12870000	27225	6.084E+09			
185	90000	16650000	34225	8.1E+09			
185	92000	17020000	34225	8.464E+09			
195	93000	18135000	38025	8.649E+09			
210	108000	22680000	44100	1.166E+10			
1047	519000	93561000	189249	4.633E+10			

TENSILE STRENGTH versus ROCKWELL HARDNESS "B" SCALE

Wrought copper alloy C69000 (alloying elements-zinc, aluminum, and nickel)
Hardness range (79-99) HRB

$$\sigma = 2137 \times HRB - 91620$$

Rockwell hardness	Tensile strength, psi	*XY*	X^2	Y^2	Parameters of the regression line		Correlation coefficient
HRB, X	***σ, Y***				***slope***	***intercept***	
79	82000	6478000	6241	6.724E+09	2137	-91620	0.919
90.5	94000	8507000	8190.25	8.836E+09			
95	105000	9975000	9025	1.103E+10			
97	113000	10961000	9409	1.277E+10			
98	120000	11760000	9604	1.44E+10			
99	130000	12870000	9801	1.69E+10			
558.5	644000	60551000	52270.25	7.065E+10			

TENSILE STRENGTH versus BRINELL HARDNESS

Wrought copper alloy C69000 (alloying elements-zinc, aluminum, and nickel)
Hardness range (150-240) HB measured at 3000-kgf load

$$\sigma = 506 \times HB + 3520$$

Brinell hardness	Tensile strength, psi	*XY*	X^2	Y^2	Parameters of the regression line		Correlation coefficient
HB, X	***σ, Y***				***slope***	***intercept***	
147	82000	12054000	21609	6.724E+09	506	3520	0.964
187	94000	17578000	34969	8.836E+09			
210	105000	22050000	44100	1.103E+10			
222	113000	25086000	49284	1.277E+10			
228	120000	27360000	51984	1.44E+10			
237	130000	30810000	56169	1.69E+10			
1231	644000	134938000	258115	7.065E+10			

Appendix XII

Hardness and Tensile Strength Relationship for Cast Copper Alloys

TENSILE STRENGTH versus BRINELL HARDNESS

Cast copper alloys, C8xxxx series (alloying elements-zinc, tin, and lead)
Hardness range (50-80) HB measured at 500-kgf load

$$\sigma = 753 \times HB - 8220$$

Brinell hardness *HB, X*	Tensile strength, psi *σ, Y*	*XY*	X^2	Y^2	Parameters of the regression line *slope*	*intercept*	Correlation coefficient
53	34000	1802000	2809	1.156E+09	753	-8220	0.926
55	34000	1870000	3025	1.156E+09			
59	38000	2242000	3481	1.444E+09			
60	35000	2100000	3600	1.225E+09			
62	35000	2170000	3844	1.225E+09			
62	37000	2294000	3844	1.369E+09			
76	51000	3876000	5776	2.601E+09			
427	264000	16354000	26379	1.018E+10			

TENSILE STRENGTH versus BRINELL HARDNESS

Cast copper alloys, C86xxx series (major alloying elements-zinc and aluminum)
Hardness range (110-225) HB measured at 3000-kgf load

$$\sigma = 472 \times HB + 11890$$

Brinell hardness	Tensile strength, psi	*XY*	X^2	Y^2	Parameters of the regression line		Correlation coefficient
HB, X	***σ, Y***				***slope***	***intercept***	
108	65000	7020000	11664	4.225E+09	472	11890	0.997
130	71000	9230000	16900	5.041E+09			
155	85000	13175000	24025	7.225E+09			
180	96000	17280000	32400	9.216E+09			
225	119000	26775000	50625	1.416E+10			
798	436000	73480000	135614	3.987E+10			

TENSILE STRENGTH versus BRINELL HARDNESS

Cast copper alloys, C87xxx series (major alloying elements-zinc and silicon)
Hardness range (70-120) HB measured at 500-kgf load

$$\sigma = 265 \times HB + 35910$$

Brinell hardness	Tensile strength, psi	*XY*	X^2	Y^2	Parameters of the regression line		Correlation coefficient
HB, X	***σ, Y***				***slope***	***intercept***	
70	55000	3850000	4900	3.025E+09	265	35910	0.982
85	58000	4930000	7225	3.364E+09			
85	58000	4930000	7225	3.364E+09			
115	68000	7820000	13225	4.624E+09			
118	66000	7788000	13924	4.356E+09			
473	305000	29318000	46499	1.873E+10			

TENSILE STRENGTH versus BRINELL HARDNESS

Cast copper alloys, C92xxx and C93xxx series (alloying elements-tin, lead, and zinc)
Hardness range (50-75) HB measured at 500-kgf load

$$\sigma = 761 \times HB - 10930$$

Brinell hardness	Tensile strength, psi	*XY*	X^2	Y^2	Parameters of the regression line		Correlation coefficient
HB, X	***σ, Y***				***slope***	***intercept***	
48	27000	1296000	2304	729000000	761	-10930	0.952
58	32000	1856000	3364	1.024E+09			
60	32000	1920000	3600	1.024E+09			
64	41000	2624000	4096	1.681E+09			
67	38000	2546000	4489	1.444E+09			
67	39000	2613000	4489	1.521E+09			
70	42000	2940000	4900	1.764E+09			
70	45000	3150000	4900	2.025E+09			
72	44000	3168000	5184	1.936E+09			
75	46000	3450000	5625	2.116E+09			
651	386000	25563000	42951	1.526E+10			

TENSILE STRENGTH versus BRINELL HARDNESS

Cast copper alloys, C94xxx, C96xxx, and C97xxx series
(major alloying elements-nickel and tin)
Hardness range (60-175) HB measured at 500-kgf load

$$\sigma = 457 \times HB + 8220$$

Brinell hardness	Tensile strength, psi				Parameters of the regression line		Correlation coefficient
HB, X	***σ, Y***	***XY***	***X²***	***Y²***	***slope***	***intercept***	
60	36000	2160000	3600	1.296E+09	457	8220	0.989
80	45000	3600000	6400	2.025E+09			
85	47000	3995000	7225	2.209E+09			
85	50000	4250000	7225	2.5E+09			
114	55000	6270000	12996	3.025E+09			
175	90000	15750000	30625	8.1E+09			
599	323000	36025000	68071	1.916E+10			

TENSILE STRENGTH versus BRINELL HARDNESS

Cast copper alloys, C95xxx series
(major alloying elements-aluminum and nickel)
Hardness range (140-250) HB measured at 3000-kgf load

$$\sigma = 439 \times HB + 16860$$

Brinell hardness	Tensile strength, psi	*XY*	X^2	Y^2	Parameters of the regression line		Correlation coefficient
HB, X	*σ, Y*				***slope***	***intercept***	
140	75000	10500000	19600	5.625E+09	439	16860	0.961
140	75000	10500000	19600	5.625E+09			
160	96000	15360000	25600	9.216E+09			
170	90000	15300000	28900	8.1E+09			
170	90000	15300000	28900	8.1E+09			
180	95000	17100000	32400	9.025E+09			
195	110000	21450000	38025	1.21E+10			
200	102000	20400000	40000	1.04E+10			
225	116000	26100000	50625	1.346E+10			
248	123000	30504000	61504	1.513E+10			
1828	972000	182514000	345154	9.678E+10			

REFERENCES

1. *Metals Handbook, 9th Edition, Volume 8: Mechanical Testing*. Materials Park, OH: ASM International, 1985, pages 69-113.

2. *Machinery's Handbook, 25th Edition*. Edited by Robert E. Green, et al. New York: Industrial Press Inc., 1996.

3. Wadsworth, Harrison M. *Handbook of Statistical Methods for Engineers and Scientists*. New York: McGraw-Hill, 1990, pages 13.1-13.11.

4. *Engineering Properties of Steel*. Edited by Philip Harvey. Materials Park, OH: ASM International, 1998, pages 7-462.

5. *Metals Handbook, 10th Edition, Volume 1: Properties and Selection: Irons, Steels, and High-Performance Alloys*. Materials Park, OH: ASM International, 1990, pages 20, 767, 864.

6. Roberts, G. A., and R. A. Cary. *Tool Steels, 4th Edition*. Materials Park, OH: ASM International, 1985, pages 364, 365.

7. *ASM Handbook, Volume 2, Properties and Selection: Nonferrous Alloys and Special-Purpose Materials*. Materials Park, OH: ASM International, 1992, pages 49-51, 617.

8. *Titanium and Titanium Alloys, Source Book*. Materials Park, OH: ASM International, 1982, pages 10-19, 211-220.

9. Castner, M. "Turning Stainless Made Painless," *Cutting Tool Engineering*, (March 1997).